CRETACEOUS
DAWN

L. M. Graziano
and
M. S. A. Graziano

Leapfrog Press
Teaticket, Massachusetts

Cretaceous Dawn © 2008 by L. M. Graziano and M. S. A. Graziano
All rights reserved under International and Pan-American
Copyright Conventions

Published in 2008 in the United States by
Leapfrog Press LLC
PO Box 2110
Teaticket, MA 02536
www.leapfrogpress.com

Distributed in the United States by
Consortium Book Sales and Distribution
St. Paul, Minnesota 55114
www.cbsd.com

First Edition

Library of Congress Cataloging-in-Publication Data

Graziano, Lisa M.
  Cretaceous dawn : a novel / L.M. Graziano and M.S.A. Graziano.
    p. cm.
  ISBN 978-0-9815148-3-3
  1. Time travel--Fiction. 2. Dinosaurs--Fiction. I.
Graziano, Michael S. A., 1967- II. Title.
  PS3607.R39934C74 2008
  813'.6--dc22
      2008013975

Printed in the United States of America

"This is a book that I read nonstop from start to finish . . . a gripping adventure about three scientists' journey back in time 65 million years to the Late Cretaceous of South Dakota. As a paleontologist who studies these animals and ecosystems, I found in *Cretaceous Dawn* wonderful and accurate images of what it must have been like in that long-gone world. This book pulls you into the Cretaceous where you can feel being scratched by the dense forest vegetation, taste the meat of the dinosaurs the characters are eating, and feel your sweat from living in a tropical, Greenhouse climate. Of course, the most important difference 65 million years ago was survival in a world full of fascinating and terrifying animals such as raptors, *Tyrannosaurus rex*, and *Triceratops*. A must read for anyone curious about dinosaurs—and who isn't?"

—Julia Sankey, Ph.D., Vertebrate Paleontologist

The genuine science in *Cretaceous Dawn* is rendered with clarity and vividness and gives the novel its richness. *Cretaceous Dawn* is plain fun, and educational at that. Short of time travel, this is as close as you'll ever get to the grim, silent, predatory world of the Cretaceous.

—*Falmouth Enterprise*

A fresh take on a classic idea, a solid imaginative feat, and a treat on many levels. A page-turner, starting off fast and never letting up. The dinosaurs and other prehistoric goodies are alive: specific, detailed, extraordinarily available to the senses.

—Mary Patterson Thornburg, author of *Underland*

A riveting book, and the accuracy was spot-on.

—Lorin King, Curator, Dinosaur Depot Museum

A real page-turner, absorbing and exhilarating enough to thrill any reader.

—Jon P. Stone, Executive Director, Dinosaur Depot Museum

[*Cretaceous Dawn* is] a thriller that combines paleontology, physics, and a missing-persons case. The Grazianos have created a lively reading experience set today and 65 million years ago. Among their sources for this gripping story are a real-life professor of vertebrate paleontology and a curator at the Denver Museum of Nature and Science.
—Cape Cod Times

For T. E. and all other dinosaur enthusiasts

ACKNOWLEDGMENTS

We gratefully acknowledge our editors, Kim Davis and Jessica Buchingham, for their help through several drafts. Dr. Julia Sankey, professor of vertebrate paleontology at California State University, enthusiastically reviewed the manuscript for scientific accuracy, and Dr. Kirk Johnson, chief curator at the Denver Museum of Nature and Science, kindly reviewed the information on Cretaceous vegetation.

– Prologue –

Yariko Miyakara was a crack young physicist, but at the moment she had a very annoyed expression on her face as she stuck her head out of a small round doorway in the back wall of the lab.

"There's another one!" she cried.

"You're kidding." Yariko's graduate student, Mark Reng, was sitting at a computer in the main lab. Mark was only twenty-five, a new student for Yariko, but he was quick to learn and full of odd ideas, and he loved the lab. His unruly mop of brown curls could be seen bobbing about the rooms at all hours of the day and night as he puttered with small experiments and computer programs. Now he came over to the vault and looked in past Yariko. "The same kind?"

"Who knows. Get me another jar." When Yariko climbed out of the vault she held the glass jar in one hand. The other hand was tugging on her long braid, a habit she had when something didn't make sense.

Mark took the jar from her. "I don't understand," he said. "I cleaned the vault. First thing today. There wasn't a speck of dust. And certainly no . . . bugs." His face showed disgust as he glared at the thing inside the glass.

A rather large, red-and-gold beetle calmly waved its antennae at him, and began an exploratory climb up the side of the jar. "How does a beetle get into a lead-lined, sealed graviton vault in the middle of a physics building?"

"Not by itself, that I know." Yariko took the jar back. "I'll put it on the shelf with the other two. I find this. . . ." She stared at the insect as if it might explain its own mysterious appearance in the graviton vault. "Provoking," she finished.

Provoking indeed. Annoying, perplexing, and perhaps . . . exciting? Certainly, beetles were not created inside a sealed vault. And so, they had come from outside. They had been brought in as a by-product of the experiment. There was a word for this: translocation. But Yariko did not even say the word in her mind, not yet, not until she was sure.

Mark shook his head and turned back to the computer. "Provoking is right. Someone's playing a bad joke on us. Do you think they bite?"

"Who knows." Yariko turned away and unlocked a door, revealing a small chemical storeroom, where she carefully placed the jar on a high, dusty shelf. Two jars already stood there, old condiment jars with holes poked in the lids. A similar-looking beetle scurried around inside one; in the other, a small brown insect lay still. "It's not much of a joke," she said quietly to herself.

"I'm off," Mark called from the outer room. "Homework session. Tensor equations today—try explaining that to a bunch of undergrads."

Yariko waved in acknowledgment. "I'm off too. Faculty meeting. We'll see what happens when we start this afternoon's run."

"Just don't make any more beetles," Mark said with a grin.

Yariko was thoughtful that morning, abstracted, hardly paying attention to the meeting's discussion. She didn't miss much: faculty meetings were notoriously boring, cluttered with petty gripes about bulletin board space and graduate students blasting music in the labs. It was much more interesting to think about beetles.

When she returned to the lab two hours later, she made an entry in the notebook and then checked some lines of program on the computer. She tapped her pen absently on the keyboard as equations rolled by in her mind. What should the next experimental run look like? Was it worth fiddling with the parameters?

The beetles were exasperating. Impossible. There were no beetles in the vault before the run started. There was no way they could have gotten in. In effect, her analytical thought said, they could not exist.

Yariko suddenly threw down her pen and strode to the supply room. It had automatically locked behind her, earlier; a required safety feature that annoyed her each time. She found the small key on her chain and opened the door.

The three little jars stood on their high shelf in the gloom of a badly lit corner. She lifted the newest one down.

It was empty. They were all empty.

Part I

Cypress Island

The coleoptera, the humble beetles, are by some measures the most successful type of animal on the planet, comprising about 300,000 known species. They were just as abundant in the Cretaceous. If you could magically reach back to that era, pick one animal at random, and bring it to the present, odds are that you would retrieve some kind of beetle.
—*Julian Whitney*, Lectures on Cretaceous Ecology

– ONE –

Julian Whitney leaned back to enjoy his last quiet hour that Monday morning, propped his feet up on the desk, and spread the latest *Paleontological Progress* on his lap. Autumn had just arrived at the tiny university of Creekbend, South Dakota. Classes had begun, and that afternoon he was scheduled to give the first of his lectures on Cretaceous Ecology.

As a paleontologist, especially a young one active in field work, Julian was comfortably relaxed in regards to dress: his worn jeans were dusted with fine soil from the lab, the once-white sneakers showed glimpses of black socks through small acid holes in the toes, and his brown hair, untrimmed for too long, flopped over his forehead. As he flipped through the articles in the *Progress* he unconsciously felt for the small compass bulge in his watch pocket.

The phone shattered his little moment of peace, and he let it ring a few times before giving in and answering it.

"Julian," came the voice at the other end, without preamble, "are you free? Can you come over here and look at something?"

He recognized the voice of Yariko, a striking-looking physicist with a formidable reputation. He'd met her several times at faculty gatherings, and had found her friendly but somehow unapproachable at the same time. Now here she was sounding strangely excited, asking him to come to her lab; and before he could answer, she asked a very strange question for a physicist. "What do you know about insects? Beetles?"

He took his feet off the desk and sat up. "Beetles? I only know about beetles that are really, really old. I'm hardly an expert on living ones. If you talk to Bob Heckwood. . . ." Even while speaking Julian cursed himself for the habit of self-effacement that made him put forward someone else.

"Never mind Bob Heckwood. Just get over here."

Another voice could be heard in the background, and Yariko answered cryptically, "Startup dot sixty-three. It's the newest version of the program. He'll come. Julian," she said into the phone again, "be quick. It's a small one, so it won't last long." She hung up.

Julian sat still for a moment, puzzling over her strange statement. Beetles? Why was Yariko interested in beetles? And why the hurry? It would still be there even if it died, or didn't "last long" as she put it. He hesitated, unsure how he could help, wondering if he should call Bob Heckwood and send him over. But it wasn't a serious thought.

He stood in sudden decision and scanned the bookcase for a ratty old Audubon Guide to North American Insects, which he knew was there. He cursed it for evading him, then found it turned wrong-way on the shelf. He threw it into his "briefcase," really a paper grocery bag, threw in a notepad and a pen, and hurried out.

The campus was busy with students moving between classes, chattering students in groups with bulging backpacks slung on their shoulders. A few greeted Julian as he hurried past, and one tried to stop him with a question, but he hardly noticed.

Julian had to admit he was flattered by the strange phone call. It was not only her reputation; it was that she had noticed him after all. He thought back to their first meeting at last year's Christmas party, shortly after she'd been hired. He was immediately drawn to her: her long black hair was draped over one shoulder, she wore no makeup or jewelry, and the elegant simplicity of her slender form was stunning. He'd watched her covertly for some time, wondering who she was, watching her smile and converse with a seeming ease that he never felt in such gatherings.

Unfortunately, their own introduction did nothing to put him at ease. When she said "particle physics" in response to his question,

Julian was intimidated and a bit at a loss for conversation. Even now, walking across campus on her invitation, he tried not to remember how he'd babbled about fossilized pollen while the humor grew in her eyes, until, mercifully, a colleague had whisked her away for more introductions.

True, she had always been friendly since then, and once they'd even spent an hour talking quite comfortably together, after she'd given an informal performance, at the campus coffeehouse, on a strange five-stringed Japanese instrument. There were a few times like that when she was less aloof; other times when they crossed paths she was abstracted, distant, and Julian imagined physics equations tumbling around in her head unceasing while her eyes looked at him.

As he reached the physics building the morning began to play itself out in his head: he would give Yariko and her colleagues critical information that would result in a publication, and they'd work together in the future on some strange experiment. . . .

Who ever heard of a physicist interested in beetles?

The physics department was full of dark corridors, gritty, windowless, lit by a few feeble bulbs hanging from the ceiling. Following the signs saying "Graviton Lab," Julian clattered down the old rusty fire stairs to the basement and hurried past a few turns of the cinderblock passageway. The door was closed, and he was surprised to see a sign taped up that said:

GRAVITON LABORATORY
PROJECT 354S
DO NOT ENTER
AUTHORIZED PERSONNEL ONLY

He'd never been in this part of the building before, and was surprised to see anything so forbidding. Just as shocking was the keypad mounted on the wall beside the door, apparently an alarm system, but it looked to be unarmed since none of the warning lights were illuminated.

Yariko's voice drifted back when he knocked on the door. "Come in, Jules!"

He opened the door a crack and peered inside. The front room of the lab was a mess of papers, Xeroxed articles, strange rune-like printouts from various experiments, books and journals scattered all over a work table and mixed up with a few grimy coffee cups. Yariko was not in sight; she must have been in the main room tending to an experiment.

Julian hesitated. "Am I authorized?" he called out.

"Just come in." Yariko's voice, still impatient, came from a doorway to the right. "What took you so long?"

"I was looking for a book," he said, following her voice through an inner door and into the gloom of Experimental Setup A. The lighting was poor because most of the fluorescent tubes in the ceiling had burnt out. Typical of any science, he thought: in the midst of cutting-edge technology you could always find something very basic that didn't work.

"Why are you standing in the doorway?" Yariko said, and Julian realized he was curling and uncurling the top of his paper-bag briefcase as he stared around the lab.

Yariko gave him one of her brilliant smiles. "It's all right. We don't eat paleontologists here."

He gave her a sheepish smile in return, and entered the room, still gazing around in curiosity.

There was a long counter against the wall, five squat IBMs arranged along it. They seemed to be rigged as control devices; cables sprouted out of their back panels, ran up the wall and along the ceiling, stuck up with duct tape in a messy, homemade fashion, and then disappeared into a small hole that had been drilled through the concrete high up on the opposite wall. Beneath was a circular metal door, about a yard in diameter, exactly like a safe-deposit vault. Clearly the cables fed some device inside the vault.

Yariko was sitting on a stool at one of the consoles, and as Julian approached she turned back to typing in strings of numbers with incredible speed. The other monitors displayed a typical screen saver, fanciful pictures of marine life, fish swimming slowly back and forth past green and red corals, glowing in the dim light of the room.

"*Carassius auratus*," Julian said.

"Excuse me?" Yariko said, continuing to input numbers. They apparently came out of her head.

"The common goldfish."

Yariko propped her elbow on the counter with her chin on her fist and gave him a long look that twinkled with humor. "You're too late."

"I was looking for a book," Julian repeated. "On North American insects. Why am I late? What did I miss?"

She lifted a glass specimen jar from the counter top. It looked like a mayonnaise jar with the label washed off. "This," she said, giving it a shake. The jar was empty.

"A bug escaped from you?" Julian tried to match Yariko's light humor.

"Yes, in a manner of speaking." She closed the notebook and swiveled her stool to face him. "Have a seat."

He sat down with the paper bag in his lap and asked, "What's the little round door? A Hobbit hole?"

As he spoke the door was pushed open slightly, and from behind it a man's voice could be heard, swearing quietly.

Julian stiffened. Perhaps this was *the* man, the mysterious special friend of Yariko's, who some even called a fiancé; the dratted man whose existence, discovered last spring, had caused Julian to retreat with a sigh to his familiar fossils. The fiancé lived elsewhere, apparently far away. Long-distance relationships were all too common in academia. But perhaps the fiancé was visiting now. Perhaps they were collaborating on some strange beetle-related project.

"What's the matter? It's only Dr. Shanker." Yariko waved toward the little door. "He's recalibrating the instruments for the next run."

Julian relaxed, feeling slightly foolish. He'd heard of Dr. Shanker, another Creekbend physicist and far too old to be Yariko's boyfriend. He indicated the round door and asked, "Why all the heavy armor? I've never heard of a lab with a steel safety vault."

"Explosion proof," Yariko said, as if it should have been obvious. "A safety requirement."

"And the burglar alarm?"

"ONR."

Julian looked at her blankly.

"Office of Naval Research," she said. "They want to keep our work safe and secret. The security guards are at lunch right now, or they probably wouldn't have let you in."

"Security guards?" Julian was shocked.

Yariko laughed her wonderful rich laugh. "Don't worry. They won't break your knees."

"Then tell me about these secret experiments," he said, hitching his chair a little closer and lowering his voice.

"I'll give you the short version, and skip the incriminating details. What do you know about quantum gravity?"

"Nothing."

"Don't grimace," Yariko said. "It's not that bad. The general ideas are simple enough. Two planets attract each other because they exchange gravitons, quantum particles. In theory. But the theory's a mess, and the correct mathematical form of it has never been worked out. A great sticking point is that nobody has detected gravitons. Naturally, people would like to measure these particles before believing in them; or, better yet, we'd like to produce them in the lab."

"Is that what you're working on here? Gravitons?"

"Never mind that. I didn't call you to look at gravitons. I called you because for a few weeks now we've been producing some, ah, unexpected particles." She scratched her nose with the end of a pen.

"Yes?"

"Yes. Particles like pebbles. Twigs. And beetles."

Julian laughed. "You've got beetles in the lab? Are you sure they aren't cockroaches?"

"I didn't call you here to insult us," Yariko said, and they both grinned. "The graviton vault is sealed—airtight, in fact, during the runs. Nothing gets in or out."

"But you can't create beetles out of thin air," Julian said. "There must be a nest—"

"Of course we can't just create them," Yariko interrupted. "They already exist; somehow, we're pulling them into our vault when we do these runs. What we've stumbled on," and she sat up straight and looked right at Julian, "is spatial translocation of mass."

"Spatial . . . you mean you're transporting things from somewhere else? But that's incredible!" Julian didn't know whether to laugh again or not.

Yariko nodded. "An apt word. What we'd like to know is where the things are coming from. Right under the building? Or several miles away? Or even more? We have one indication so far; but I'm hoping you can tell us more."

She turned back to the computer and opened an image file. A beetle appeared on the screen, digitally photographed in several orientations. It was garishly colored, red and gold with a hint of green in the carapace, and it had extremely long antennae.

Julian's mouth fell open. "Where did you get these pictures?" he said, leaning closer to the screen to study them better.

"I took them," Yariko said. "It was only just alive, so I was able to lay it on its back here, and get the details of the underside. Here's a close-up of its leg, and here's its little face," she added, tapping the screen with her pen.

"You took. . . ." Julian sat back and stared at her. "But, but that's impossible. This beetle doesn't exist any more."

"What do you mean it doesn't exist?"

"I mean it doesn't exist. It's extinct."

"It died out?"

"It disappeared about—"

Yariko interrupted. "But it's not. We've found several of this kind, in fact. Some arrive quite alive, I promise you."

"I tell you, this is an extinct insect," Julian insisted. "One of the archostemata, probably the Parasabatinca genus. The details on the antennae, the unique joints on the legs, the shape of the carapace . . . but how could anyone know the color? Color isn't fossilized." He looked up again from the images in the screen. "Show me this beetle."

"I can't," Yariko said. "It's gone. Disappeared. I told you, they don't persist."

Julian stood up and began to pace, suppressing a rising impatience. "Are you telling me you actually discovered a living representative of a beetle thought to be long extinct, and you let it go?"

"I didn't know it was 'long extinct,'" Yariko said, and she sounded exasperated. "I only know that it came from somewhere and was translocated through space into our vault during an experimental run to measure gravitons. I thought you might be able to tell us where it came from, and where some of the other samples came from. I want to know how far these objects are being moved."

Julian sat down again. "I'm finding this very hard to believe," he said. "And what does it have to do with your gravitons?" Somehow it was easier to believe in the spontaneous creation of gravitons, which were effectually meaningless to him, than the spontaneous creation of beetles.

"I can't tell you that," Yariko said. "It's classified."

A deep voice suddenly boomed out of the explosion-proof vault. "Classified, my ass. Tell him about it. Who are you talking to anyway? Is that your paleontologist?"

The heavy round door of the vault swung wide, and Julian blinked in surprise as a huge German shepherd leaped out of the portal.

Yariko laughed. "Hilda," she explained. "She's part of the staff. Against university policy and illegal of course, but nobody minds." The dog trotted over to Julian and thrust her nose into his lap.

Next an enormous man with a grizzled beard climbed out, carefully squeezing himself through the small opening. He came forward with his hand extended.

"Whitney, is it?" His voice filled the whole room, and his handshake was equally vigorous and overpowering. "I'm Shanker. Yorko's colleague. We've been working on this project together. Pleased to meet you. Yorko's been telling me about you, and you're just exactly the man for us. What's that? Doughnuts?" He was looking at Julian's grocery-bag briefcase, and his whole face seemed to light up at the prospect of eating something.

"A book on insects," Julian said, apologetically.

"Well, no time for doughnuts anyway. Yorko and I were just setting up for another experimental run."

Julian was startled by the name, but Dr. Shanker didn't seem to realize what he'd said. Yorko was his nickname for Yariko, as Julian soon realized; or maybe he was incapable of pronouncing it correctly.

Julian had heard the stories about Dr. Shanker, both from Yariko and from his own students who took physics classes. The man was known to teach with great flair and enthusiasm and was beloved by his students. He exercised at the gym every day and had a tremendous physique, even though he was sixty-five. The other faculty thought he was loud and coarse, and he had been married and divorced three times before finally choosing to live alone with his dog. Despite such gossip there was something hearty and straightforward about him that Julian immediately liked.

He shook Julian's hand and clapped him on the shoulder, and then he turned to Yariko and said, "What have you told him? Anything?"

"Only the by-product," she said. "The beetles and such. He's not a physicist."

"Good. You just missed the morning run, Whitney, but we're about to start up again right now. It's hard to predict how long they'll persist. A few days, some of them. Or a few minutes. Then they fizz out, or 'revert' as we hypothesize. The greater the mass, the longer they'll persist—linear function—but there's a lot of statistical fluctuation."

"He has no idea what you're talking about," Yariko said.

"Good, good," Dr. Shanker said, unperturbed. "He'll see." He turned to Julian again. "It's the triumph of the little guy," he said. "Low energy physics, on a shoestring budget. No cyclotron, no particle accelerator, no nothing, no big bucks. What do you think, Whitney, how much is the lab worth? Take a guess—what's the dollar worth of all our equipment?"

Julian had no idea. "How much?"

"Ten million," he said. "Tops."

To Julian it sounded like a staggering amount, but Shanker seemed to take it as near poverty.

"With brains," Dr. Shanker said, tapping his forehead with no sense of modesty, "with brains, there's no limit to what you can do."

Julian looked at Yariko, ready to raise his eyebrows. But she didn't return the look, and Dr. Shanker continued.

"For a while now we've been producing these by-products on almost every run of the experiment. That's where the beetles come in. We've been hesitant to make an official report, because that pack of ONR regulators would descend on us and take the experiment out of our hands. We don't want that to happen. That's why we've snuck you in here in such an unofficial fashion. Yorko says you're a good ecologist, with a sharp eye for detail, and that's what we're looking for right now. What do you think of this?"

He opened another jar and held out a small black stone.

Julian turned to the light and studied it. "This was formed at a high temperature," he said instantly. "You see the crystals? Slow cooling, especially in the middle here. I take it this is a cross section?"

Yariko nodded. "We split it and sent half to Geology for analysis. It's the only stone that didn't revert. But they all looked similar."

"What's 'revert?'" Julian asked.

"They disappear, after being moved to specific places in the lab. We think they're going back to their original location."

Julian handed the pebble back. "So that's what happened to the beetles? How come none of them stayed here?"

"Well, we don't know yet. . . ." Yariko began, but Julian interrupted.

23

"I need you to get another one right away," he said. "One of those beetles. If you've found a living specimen of Parasabatinca, if I can identify the species, why it'll be the biggest find in. . . ." he stood and began pacing again, rubbing his hands. What a coup it would be! "Can you do another run, or whatever you call it? Can I watch?"

Yariko and Dr. Shanker looked at each other. Julian stopped and pointed to the pictures still up on the computer screen. "Don't you see? This is huge. If you've actually found living specimens, if these things still exist somewhere, well it's like discovering a living dinosaur. Can't you see how exciting that would be?"

Dr. Shanker grinned and clapped Julian on the back, rather too hard. "So you're interested in our beetles. Good, good. We'll try to get one for you. Now: no more talk. You can see for yourself. Yorko, battle stations!"

Yariko took off the tattered lab coat she wore over her jeans and T-shirt and manned the computer. Dr. Shanker climbed back into the vault. "Just to check the alignment," he said. Hilda leaped through the portal after him.

Julian stood outside and peered into the vault. It was a small chamber, perfectly cubical, maybe ten feet to a side, and the walls were entirely metal. Lead lined, he was told later, as well as lined with a faraday cage to keep out radio signals and other static. It was lit by floodlights that were set into circular pits in the ceiling and screened off by the faraday cage.

There was a low shock-proof table in the center of the vault, and bolted to the table top were various instruments and canisters, a little bit of copper tubing connecting mysterious pressure chambers, a thicket of gauges and dials, and electrical wires, green, red, blue, and white, that had been meticulously soldered in place by hand.

A great twisting blue and red cable came down from a hole in the ceiling and fed the equipment. Yariko had once told him how she'd set it all up over the past few months: one of the reasons she worked at night as well as by day, since the delicate adjustments necessary were easily disturbed by noise and vibrations. Dr. Shanker was now kneeling beside the table holding a jeweler's screwdriver in his massive hand, adjusting a calibration screw and watching one of the dials with the intensity of a true and dedicated scientist.

Julian felt a strange excitement. If true, this would be the oddest discovery he'd ever made. He imagined the headlines: "Paleontologist discovers living fossil of long-extinct coleoptera in a physics lab." He grinned.

Finally Dr. Shanker stood up. He turned, with his finger to his lips, and leaned out of the portal. "No talking," he whispered. "The vibrations will throw it off again. We don't want to produce the graviton bomb here—although ONR might be intrigued if we were all killed in an explosion." He leaned farther out and whispered to Yariko, "Have you run the start-up routine?"

"Yes," she whispered. "Get out of there. I'm about to execute."

"Try a level six perturbation."

"Level six?" Yariko paused, her fingers on the keyboard.

"Too high, you think? It should increase the probability of a good result. And Whitney's come all the way over here just to witness the impossible."

She grinned. "All right. Get out, close the door."

"Don't be ridiculous," Shanker said. "Nobody will know. I want to watch and monitor the result."

"No," Yariko said. "Get out of there. Never mind that it's illegal—it's dangerous, at such a high energy level."

It sounded like a long-standing argument between them. Julian wondered why it was dangerous, but he didn't dare interrupt to ask.

"If you insist," Dr. Shanker said, "I'll close the door." He retreated inside the vault, and coolly reached out and pulled the door shut on himself.

Yariko looked up at Julian and shook her head. "He's going to get hurt someday, but it won't be my fault. I've warned him plenty of times. Do you want to step out?"

"Step out?"

"The risk is small, but not zero. Graviton vaults have been known to explode, and this one is wound up tight with these settings."

"If you're staying, I'm staying," Julian said. "I'm not missing this. I want to see that beetle." He might have to scramble to prepare for his one o'clock class, but he wasn't going to miss this for anything.

"OK then." Yariko turned back to her console, typed a few commands, and then paused with her finger hovering over the return key. "Here goes," she said, and pressed it.

A string of numbers began to scroll up the screen, too fast to read.

The other five computers on the counter top were slave machines, running routines, displaying numbers and status messages, none of which Julian understood. There was a clicking sound through the door of the vault, which was still open a crack. Julian looked at his watch and counted off a minute—a tense one.

The clicking began to speed up, like a mechanical toy running too fast; and then suddenly it stopped.

There was a moment of silence, and the door of the vault swung open. Hilda poked her head over the lip of the portal, and then Dr. Shanker looked out. He was grinning; he stuck out his big hairy fist with his thumb jutting up.

"So far so good," he whispered. "Keep it quiet still. We don't want anything vibrating while it powers down." His head disappeared again into the vault.

"Did it work?" Yariko whispered eagerly, getting out of her chair and stepping forward on tiptoe. "What are the readings?"

But at that moment a door slammed in the next room, and it sounded like a cannon shot bursting into the silence. The security guards had returned early from lunch, and could be heard talking and laughing in the front room of the lab.

Yariko's expression changed to alarm at the noise. The door of the vault should have been closed, sealing out sound and other vibrations; but as in any lab, exact protocols were followed only during inspection visits. With the guards off, and students still in class, safety measures were more relaxed.

A few seconds passed. Yariko's scowl relaxed into a wry smile. She shook her head and whispered something before returning to the computer. The guards could be heard leaving the front room again. Julian opened his mouth to ask what exactly the experiment was, and if it had created another beetle.

But Yariko was staring at the computer screen with a frightened expression on her face. She hit a few keys, apparently without effect, and then she got up and ran toward the vault.

Julian started to ask what was wrong but he never got the words out. Nor did he make it to class that afternoon.

26

Science is by definition an iterative process, and scientists, those ob-servers of the natural world, are drawn to the philosophy by their in-nate curiosity. The world view of science is changeable, constantly re-viewable, with never-ending caveats, complexities, and surprises: it is, in fact, a coarse reflection of the natural world. And therein lies the fascination.
—*Julian Whitney,* Lectures on Cretaceous Ecology

– TWO –

The explosion was deafening in the confined, concrete-walled room: a sharp crack like a massive electrical discharge. Julian in-stinctively pulled into a crouch with his arms over his head. There was a flash of blue light, and the circular door of the vault blew wide open and slammed against the wall. Smoke poured from the aperture, and the room filled up with the acrid smell of burnt elec-trical circuits.

Yariko was the first to recover from the shock; she darted to the opening and leaned in, fanning aside the smoke with her hand. "Dr. Shanker," she hissed. "Are you all right? Can you hear me?" To Julian's horror, she climbed inside.

Even in the first instant of fear, Julian noted that Yariko did not yell; if anything she was almost whispering. His ears, recovering from the explosion, barely registered her voice. For several more seconds he didn't move, not knowing what would happen next. There might be another explosion. The vault might be on fire. He wished Yariko would get out.

He didn't want to think about Dr. Shanker, in the center of the explosion. He felt slightly sick.

"Julian." It was Yariko's voice, barely reaching him. She was look-ing out of the vault, her eyes already red from the smoke. "Come here. I need your help."

Realizing he'd been crouching half under the counter while Yariko charged into the smoke, Julian hurried over and peered in after her, dreading to see blood everywhere. The smoke was already dispersing, although the smell was choking and it stung his eyes. The equipment on the center table was partially burnt. Some of the dials had been torn off and lay scattered about the room, their glass faces glittering in shards on the floor. Hilda cringed in the corner with her tail between her legs. Dr. Shanker knelt, clutching one side of his face, blood congealing in his beard. His hands were trembling.

Yariko knelt beside him. "We have to get him out," she whispered, looking up at Julian as he leaned in the doorway. "Before it explodes again. It's still sensitive to vibrations—to noise."

Julian climbed in, squinting against the smoke, trying to ignore the nauseating feeling in his stomach. They put their hands under Dr. Shanker's elbows and heaved.

Then a voice bellowed, "Don't move him if he's hurt!" A man's head was thrust into the vault: one of the security guards, Julian realized; a big, meaty fellow with a marine style crew cut and a no-nonsense air. He frowned at Julian and said, "Who are you? Who let you in?"

"There's no time," Yariko snapped, although still speaking quietly. "We have to get out of here now. And keep your voice down." To Julian she said, "Lift him up." They tugged again at Dr. Shanker's arms, without much result, as he weighed over two hundred pounds.

The guard eyed Yariko suspiciously. He climbed inside and knelt in front of Dr. Shanker. "Show us the injury," he said, and his voice, although calm, reverberated in the metal chamber. He firmly pulled aside Dr. Shanker's hands, revealing bloody pulp where his face should have been.

The guard winced. "Christ," he muttered. "All right, let's lift him out." Then he shouted, "Ron! Hurry up! Ron! I need help!" His voice echoed and boomed in the vault, and Julian heard Dr. Shanker whispering through his clutched hands and his labored breathing, "Shut up. Shut up."

Yariko stood and grabbed the man's arm. "You'll set it off again," she hissed. He stared at her as she went back to Dr. Shanker and tried to lift him again. "Help me. Julian. Frank. Just get him out of here."

The second guard leaned into the portal.

"Jesus Christ, Frank, what happened?" he said, his voice also sounding overloud in the confined space. "Electrical fire? I just called 911 . . . that was some explosion."

But then the world heaved and disappeared in blackness and noise, sparks and confusion, and a total numbness that was more frightening than any pain.

Julian struck something solid on his hands and knees; a tilted slab of concrete, he thought. But he could not get to his feet. He struggled a moment and then sank down again, cheek against the cool pebbly cement, and lost consciousness.

Once you have found a bed of fossils, how do you determine its age? Radiometric dating is not always possible, but there are a number of other methods. Ammonites are the fabulous chronometers of the dinosaur era. They lived in shallow, warm seas, and have left us the fossilized impressions of their shells in uncountable thousands. We know so much about Ammonites and how they evolved over time that merely by glancing at a few shell patterns we can date the fossil bed to within a million years.
—*Julian Whitney*, Lectures on Cretaceous Ecology

– THREE –

1 September
11:27 AM Local Time

"There should be three."

"Clearly there aren't. Unless they're hiding in a secret compartment, there's only one here, or rather one-half." Chief of Police Sharon Earles stood and carefully removed her gloves.

"That graduate student, Mark Reng, said the two physicists were in here when the explosion happened." Sergeant Charlie Hann shook his head, not for the first time that morning.

Half a body, where there should have been three. Half of a human body. The lower half: from the waist down, blood-soaked trousers, belt intact with radio and keys, a pack of cigarettes in the back pocket. And then there were the arms, again the lower halves, lying neatly and exactly in place beside the face-down—knees-down?—body; and everything sliced cleanly as if with a giant cleaver.

But who ever heard of a giant cleaver in a physics lab?

Hann turned away from the vault. "Well, let's get the experts in." He was sickened by the burnt plastic fumes, the sharp smell of blood, and the grotesque, almost obscene partial corpse. In this small town such things were not often seen, and he heartily wished he was still at the station with his coffee and doughnut on this Monday morning.

Earles watched him leave with a mixed look annoyance at his quick departure and relief at being left alone to explore her thoughts.

Outside the lab, in the dim hallway with flickering fluorescent bulbs, Hann ducked under the orange police tape and nearly collided with a young woman.

"The area is closed," he said, before the woman could ask. "We want the hallway cleared."

"I'm here for Hilda," she said. She was a student, Hann decided; maybe a senior, but still young, shocked to see thick orange tape and a police officer in a lab corridor reeking of burnt electronics. "We're going for a walk. We always do, at this time," she added, as if those were the magic words that would open the tape line. "She's in the lab. Is—is everything OK in there?"

Hann swore, and not under his breath. Make that one half where there should be four, he thought.

Sharon Earles had been heading the Creekbend police force two years. Forty-two, tough, and the town's champion pool player, she had been on her own since the age of sixteen. As the first woman on the local police force she'd had to develop a thick skin and a decisive manner that discouraged questions. Life did not give her many breaks; what success she achieved was by sheer determination and refusal to back down. Sharon was used to being on her own, and in charge.

When Hann left the physics lab, looking green about the gills, Earles took out her cell phone and called the forensic office that covered Creekbend. She knew the body was one of two young security guards, police hopefuls, assigned to this classified project; she had helped train them, and could picture his face clearly. Hann said his name was Ron McKenzie. She dispatched someone to track down the second guard, who was probably on his lunch break, and the two senior scientists, who should have been present at the time of the accident as the only two people authorized to open the vault.

The boy might turn out to be an impediment rather than an asset in this case, she thought, her mind going back to Hann. His sense of horror was only natural, but she'd expected a twenty-eight-year-old cop to be more professional. It wasn't as if she enjoyed looking at mutilated bodies either.

She stepped around the body to the small round door of the

vault. It was open now; but ten minutes ago it had been shut so tightly that it had to be forcibly pried open. As if the inside of the vault had been a giant vacuum, she thought.

For the second time that morning she stooped carefully over the corpse and studied the inside of the door. The outside, of course, was sprayed with blood, still wet. The inside, though, was clean. With her hands behind her to be sure of touching nothing, Earles leaned into the vault. The sill was clean, too. On the floor of the small, cubical chamber, near some shattered dials on a low table, there was more blood. She crawled through the doorway and went to look.

It wasn't enough. A small spattering on the cracked face of a dial, some smears on the table, several thick drops on the floor. Not enough.

It made no sense.

It meant that the other half of the body was never inside the vault.

Sharon pocketed her cell phone and stepped out, being careful not to brush up against anything, to wait for the forensics team.

• • • • •

Julian woke up in sunlight.

For a long time he lay on his back without moving, staring into a cloudless blue sky, listening to the crying of the gulls. There was a good brine smell to the air, and now and then a breeze fluttered his hair. He lay still for some time, relaxed, almost dreaming.

Finally he sat up; the world rushed in with a sudden sharpness, and he looked around, taking it all in without comprehension. He was high up on the rocky slope of a beach. At his back was a ragged line of what looked like palm trees, their great frond-like leaves absolutely still. He blinked at them and shook his head, but they remained unchanged. Before him the high sun burned over the water, hurting his eyes. On either side were tumbled rocks choked up with sand, littered with shells and decaying seaweed. A few birds, plovers they looked like, skittered over the sand at the edge of the water and snapped up bits of food. A seagull stood on a rock nearby and looked at him, cried out, pecked a few times at a crab shell, and then suddenly flew away.

At first Julian thought he was alone. When he finally glanced down at the ground he was startled—not horrified, just innocently startled. In the sand and rocks twenty feet away lay the upper fragment of a human body. It had been severed just above the waist and the arms were only stumps, exactly as if a piece of machinery had sliced through the man while he stood. The pebbles beneath him were drenched with blood, drying now to a brown crust. It was one of the security guards: Ron. His eyes were still open, staring at nothing, curiously expressionless.

Three more bodies lay in the sand: Yariko, Dr. Shanker, and Frank, the second security guard. They lay deathly still. Peaceful. The guard had an unnatural twist to his leg. Yariko's T-shirt stirred briefly in the wind.

For a moment more Julian sat hugging his knees, squinting in the intense sunlight at the vast lonely beach all around, too stunned to move. He was able to remember the details of what had happened in the lab, but at first he could not convince himself that any of it had been real. The lab seemed remote from the brightness and the warmth on the beach, like a black-and-white movie without sound, seen once long ago.

"Whitney."

Julian turned, startled. Dr. Shanker was sitting up and looking around him. One side of his face was black with dried blood, the eye swollen shut; he was grimacing as if in pain. Then Julian realized that the grimace was part smile. "Just look at it!" Shanker said. "I didn't entirely believe, but now. . . ." He got slowly to his feet, breathing heavily. When he spotted the fragment of Ron's body the smile disappeared, and the bristly bearded corners of his mouth turned downward. "One of the guards," he said in a raspy voice. "Well." His face twitched, and the mask of dried blood cracked and flaked.

Julian stared at him and then, the reality finally hitting him, got to his feet and staggered over to Yariko. When he touched her shoulder she moaned and sat up, clutching her head and squeezing her eyes shut as if she were dizzy. When she opened her eyes she gazed out over the beach, and then turned to Shanker with a bewildered look.

"What—where are we? I was in the lab," she said, getting slowly to her feet with Julian's help.

"I don't know," Dr. Shanker said. "But whatever the settings were aligned to . . . Maybe now we can solve the mystery." He glanced again at the half-body and looked quickly away.

"What mystery? What alignment? How did we get here?" Julian felt dazed by the drastic change in his environment, by the heat and color and openness, by the urgent immediacy of what couldn't possibly be real. "This isn't Creekbend, is it."

"We started in the lab," Dr. Shanker said. "In the vault. Remember what we talked about yesterday? Bi-directionality."

Yariko shook her head and pulled herself free from Julian's hand, turning to look around her. Then she saw the mutilated body and her eyes went very wide.

"He was leaning into the vault," Dr. Shanker said. "Half of him came, and half. . . . Too late now. Let's worry about ourselves first." He did not look back at the body.

"He's dead," Yariko cried.

"Quite," Dr. Shanker said, amicably. "The rest of us aren't."

Before anyone could reply he knelt beside the guard, who was just beginning to wake up. The man tried to sit but gave a gasp of pain and fell back again.

"Frank, is it?" Dr. Shanker said. The man nodded, sweating from the pain and breathing hard. "Bad luck, Frank. Broke your leg. Your calf has a pretty good kink in it."

Shanker and Yariko linked their arms around Frank and lifted him, with Julian gently supporting the broken leg. He raised his hand to block the sun from his eyes and managed not to cry out as they moved him. They carried him up the rocky slope of the beach to the fringes of the woods, and tried to lay him flat in the shade of the palm trees; but he insisted on sitting upright.

"We need splints," Dr. Shanker said, kneeling and expertly touching the injured leg. "Two good straight pieces of hardwood. This is what they teach you in summer camp; emergency first aid, in a jungle." He looked up at the others and winked his good eye. "Didn't your Daddy ever send you to camp?"

Julian found a straight piece of driftwood nearby that he broke over his knee into two lengths. He himself was beginning to wake up more and shake off the strange lassitude; although in the damp, clinging heat, he selfishly longed to collapse in the shade and sleep.

They unbelted the guard's holster and lay it aside. Using a penknife, they cut open the pant leg and exposed the purple and swollen flesh. Julian sighed in relief: the bone had not broken through the skin; it was a simple fracture. The splint was quickly accomplished with strips of fabric and driftwood. The guard moaned a few times as the ties were pulled tight, but the proper set to his leg seemed to reduce the pain. His eyes began to dart about, for the first time taking in the trees and the brown parchment-like palm leaves scattered on the sand about him.

"Where are we? Why did you bring me here? What's going on?" he hitched himself up straighter and glared around at the beach, and then at his companions. His face was alternately red and pale, and he was dripping with sweat.

Yariko and Dr. Shanker looked at each other. Finally Yariko said, "There's been an accident. . . ."

When she didn't elaborate, Julian said, "So, you think we've been sent somewhere by your vault." Suddenly he remembered the strange beetle on the computer screen, in all its garish color. "Maybe we've gone to wherever those beetles came from? Is that what you're saying? But is that possible?"

Yariko shrugged, "It's as possible as what we were seeing in the lab. That's my guess, anyway. I don't see any sign of the lab here, and like you said, we're not in Creekbend anymore."

Julian glanced again at the shining beach and the calm, bright water, at the line of palm trees at their backs, and, now that he was more alert, at the thick growth of what looked like laurel under the palms. Tropical vegetation, like that he'd seen once on a trip to Honduras. They must be very far from South Dakota.

"Where's Ron?" the guard said, breaking in on Julian's thoughts.

Nobody spoke.

Finally Yariko said, quietly, "I'll stay here with Frank. You two bring the body out of the sunlight, or cover it with sand. We can at least bury it."

Frank stared at her, and then cursed.

Julian would have liked to stay in the cool of the woods sitting with Yariko, as far as possible from the horrible corpse, but he sighed and followed Dr. Shanker. The curious detachment still had not left him. It was as if he was observing someone from a distance. It didn't matter what they did; none of it could possibly be real.

They came out of the woods into the bright glare and stepped over the stones to the site where they'd first woken up.

The body was gone.

The rocks and sand were stained with blood, marking the exact location, but the body was gone.

Between where it had lain and the water a sinuous track had been left in the sand.

"It's been dragged away," Dr. Shanker said. "Toward the water." He stooped to look closer. "Whitney. How on earth?"

Julian shook his head. His first thought was that a boat crew might have picked up the body and then left without searching for survivors. He gazed over the water, shading his eyes against the late morning sun, but the flat sheet of the sea was empty. A few birds, gull-like but with strangely long beaks, swooped to catch fish far out over the surface. The sky was clear, a saturated blue, and aside from the restless glinting of the sunlight on the waves, the visibility was perfect. They would have spotted a boat, even a small one. And why would a boat crew drag the body through the sand? "Let's follow the trail," he said, straightening up.

It led down the rocky slope and then meandered along the beach before finally disappearing into the water. To either side of the furrow were marks like footprints, but the sand was fine and dry, allowing no clear impression. Julian thought there must have been two things, one on either side, pulling the body along.

At the very edge of the water he saw the first clear marks. Huge footprints were pressed into the wet sand, and Julian realized with a shock that they all belonged to one animal, a four-footed one, obviously straddling the corpse. But there were no returning footprints. "I know that spoor," he muttered.

Dr. Shanker looked at him, waiting.

"Crocodile," Julian said. "It must have been sunning nearby, and we didn't see it against the fallen palm leaves. As soon as we left, it ran in and snatched what it wanted."

"We won't have to deal with the body, anyway," Dr. Shanker said.

Julian ignored the crude comment. Despite his pat explanation, he was still confused. The size of the tracks was astonishing. If this were one animal, based on the separation between the left and right prints it was about six feet wide and therefore more than thirty feet in length. But something else bothered him, although he could not quite place it. Something was subtly wrong.

Julian crouched to inspect a seashell pressed into one of the footprints. The spiral shape of it was so familiar that for a moment he did not even realize the terrible strangeness of finding it tossed up here on a beach. He picked it out of the sand, brushed it off, and turned it a few times in his hand.

As the truth struggled into his head, he reached out against the sand to steady himself. Faintly through a ringing in his ears he heard Dr. Shanker saying, "What is it? Whitney, what do you see?"

"It's not possible," Julian muttered. It was all he could say for a moment. Finally he looked up at Dr. Shanker and held out the little seashell. It was a flat spiral, about two inches across. Part of the shell had been crushed, and it was this damage that had allowed him to see the critical details of the suture line. He had seen this type of shell only in textbooks, and embedded in rock. He had never seen the actual beautiful pale pink color, the delicate flecks of sky blue. He had never seen a fresh one.

"What is it?" Dr. Shanker said.

"Ammonite." Julian did not trust himself to stand up yet. Instead, he sifted through the sand for more shells and found another beautiful example, larger than the first and unbroken.

Dr. Shanker crouched down on the sand. "What's ammonite?"

"Cephalopod. Mollusk. Ammonites are mollusks." Julian was hardly paying attention.

"What do you find surprising about them?"

"They're extinct."

Dr. Shanker's head came up with a snap. "Ridiculous," he said. "There must be a mistake. They died out recently, and a colony of them still survives somewhere."

"No." Julian let the sand fall through his fingers. "Not recently."

"How long ago?"

Finally Julian looked up, and their eyes met. "Sixty-five million years," he said.

Whatever Dr. Shaker had expected, it was not that. His face twitched, and his swollen eyelid opened a slit. "How long?" he said.

"Sixty-five. Million."

Deinosuchus rugosas, the giant prehistoric marine crocodile, had the strongest jaws of any animal known, at any time period. Crocodilians are often thought of as slow on land, crawling on their bellies with bent legs and feet sticking out to the side. But even modern animals have preserved the unique crocodilian ankle joint that allows the foot to rotate. A foot that can point forward allows the legs to come under the body, and the animal to lift its body off the ground. Thus crocodiles have several gaits including the gallop, which is much like a rabbit's run with hind feet and then forefeet moving in tandem. Bony scales along the back make erect walking possible even in fifty-foot animals—and Deinosuchus, *which could outcompete even the largest of dinosaur carnivores, is known to have reached such massive size.*
—Julian Whitney, Lectures on Cretaceous Ecology

– FOUR –

1 September
1:11 PM Local Time

"What have we got so far?"

Sharon Earles sat in the headquarters "conference room," really a former storage room off the entranceway, with Hann and the forensics expert from Roscoe. The man looked seasoned, she noted with approval: gray-haired, solid, calm. One didn't usually see big city experts in South Dakota towns; they stayed where the action was. Well, she thought grimly, here was some action.

The man, who had introduced himself only as Agent Kayn, gave a summary. "Three separate blood samples were collected in and near the vault. Your call to the lab this morning was very helpful— my samples were top priority. I've got blood type right now, but more details will be coming later. Based just on type, one sample matches the dead security guard, as expected. The second sample does not match the other security guard, based on the medical records you gave me. I can't tell yet if it matches either of the two researchers presumed to have been present."

38

Hann shifted in his chair. "They obviously weren't present," he said. "I don't think the other guard was either." He hesitated, and then added quietly, "Frank, the other guard, is my half-brother, you know."

Earles looked up in surprise. It was news to her that Hann had such a stake in the case. So that was why he'd left the lab so quickly. And now he wanted to convince himself that Frank hadn't even been there. Why, even after seeing which guard the body belonged to, he must have been terribly upset. And Frank had yet to be located.

Agent Kayn ignored Hann. "The third sample did not come from either of the scientists or the other guard; somebody else, somebody unauthorized and not logged in, seems to have been present at the—"

"Hilda," Hann said, breaking in suddenly. "There may have been someone named Hilda. Did you check the list of students who might work for them? Family members?"

"Hilda Shanker," Earles said, "apparently had complete blood work done only last month, prior to a minor surgery The appointment was recorded in Shanker's pocket calendar, found in his jacket. I called the local doctor's office, and the regional hospital, without result. There was no such patient admitted on or within a month of that date. Furthermore, there are no records anywhere of a Hilda Shanker. As for Shanker and Miyakara, they seem to have disappeared. Not in the building, not at their homes, missed some big departmental meeting at twelve-thirty, last seen early this morning going into the lab." Earles looked at Agent Kayn.

"There's no surveillance system, and the alarm was off," he said, "though I don't suppose there's usually such security on a research lab. But there were clearly two people working in the room earlier today, and according to the computer activity, at least one was there until about twenty seconds before the explosion."

"The guard—the dead one—called that in, so he wasn't killed immediately." Hann got up and dumped half a cup of Cremora into his instant coffee. "The real explosion was about one minute later. Shook the whole building, and presumably that's when he was caught."

The forensics man nodded. "Cause of death is obvious, although I've never seen such clean fragmentation of a body. Skin, organs, spine, everything is neatly severed. No crushing or tearing, no evidence of trauma at all in fact—except that he was cut in half. Another thing: there should have been a great deal more blood."

"Explain," Earles said, looking at him sharply, her own observation coming back to her.

"Well, what was there came from what we saw: the lower half of a body. The blood associated with the torso and head would have been an even greater volume—and it isn't there. Not in the lab, not in the vault. No blood, no torso. It's almost as if his upper half, blood and all, simply . . . vanished."

Hann didn't get his disbelieving snort out, because his boss began talking again.

"I'm calling in some physicists," she said. "I talked with that graduate student who was at the scene first. His name's Mark Reng. He's pretty shaken, but he did put me onto the top physicists in the country doing the same kind of work this lab was doing. Two of them are on their way now; colleagues of Miyakara and Shanker. Something happened in there, something that wasn't just equipment exploding. In the meantime—" Her cell phone rang. "Excuse me," she said. "This may be relevant."

The others listened in silence until the end of the call, which was brief.

"OK, we've tracked down Hilda Shanker's records." Earles looked amused. She glanced at Hann. "That student you met, the one who was 'going for a walk' with Hilda, has just been interviewed by Campus security. Don't bother," she went on, turning to Agent Kayn, whose pen was poised over his notebook. "Hilda's appointment was at the Creekbend Veterinary Hospital, which has all her medical records. I'm told she's a healthy three-year-old German shepherd."

The man threw down his pen. "A dog? A dog in a lab? Aren't labs supposed to be clean and all that?"

Earles shrugged. "A lab is like any workplace. Look at this place." The station was in fact grimy, shoddy, and smoky; stray dogs and even cats were common visitors. "There's one more thing," she went on, as Hann rose and turned to leave. "Campus security also reported another missing faculty member. Guy named Whitney, over in the geology building. Seems he never showed up for lecture, nobody could find him. Last seen leaving his office at about elcven, carrying a paper bag."

"What is this, missing persons day?" Hann asked. "Not a suicide suspect, I hope."

"They checked the phones," Earles said. "At 10:54, Whitney received a call from the particle physics lab."

• • • • •

Julian stood at the edge of the forest, where the palm trees and laurel-like bushes grew sparsely in the rocks and sand, letting in splashes of sunlight between them. The beach was quiet except for the occasional crying of the strange gull-like birds, and there was hardly any sound from deeper in the forest; only now and then the rustling of a beetle in the dead leaves, or the cawing of a bird in the canopy. The sun was high over the trees, past its zenith, and by the direction of its motion he knew they were on an eastern shore.

Looking down at his fect he saw countless shells in the sand, much as he would have seen on any beach. He felt dizzy, and he couldn't suppress the images of cool streams winding through hemlock and hardwoods, streams strewn with rocks, black rocks indented with immeasurable imprints of fossil shells. Sometimes he and his father had lost themselves in those woods. They would wander for an hour or more in search of descending ground and hemlock, indicating a stream. Sometimes, Julian imagined bears and once he was quite sure they saw a black panther, loping through the trees, eerie and agile.

But never had he been truly scared. And if he was nervous, the wonder of the streams always saved him: the streams meant rocks, and rocks meant fossils: mollusk-like shells from 400 million years ago, long before the earliest dinosaurs reared their heads in the mist. He shuddered and turned away.

Yariko and Dr. Shanker were whispering together a few feet away, stooping over a patch of sand; Yariko had a twig in her hand and seemed to be drawing lines on the ground. No doubt she's trying to figure out how we got here, Julian thought. Trying to come up with just the right equation to explain it. But what did it matter? They were here.

Frank leaned back against a palm trunk with his splinted leg jutting out straight along the ground. He must have been in great pain still; his face went pale and flushed by turns. He had unbuttoned the front of his vest in the heat and set his gun and VHF radio on the ground beside him. Somehow, he still retained the imposing authority of a policeman; he looked alert, if immobile. He was perhaps twenty-five, large and broad-shouldered. His blonde crew cut was frosted with sand stuck in the sweat. He turned to look at Yariko and Dr. Shanker. "What are you doing?" he asked, his voice sounding strained.

"Trying to figure out how we got here," Dr. Shanker replied shortly. "And where, or when, here is." He looked at Julian again.

Yariko paused in her sand writing. "Are you sure of what you saw?"

Julian felt a little bit insulted. "Of course I'm sure. Just like I was sure about the beetle—only you didn't listen. You kept insisting it wasn't extinct."

"All right, no need to get angry," Dr. Shanker interrupted. "If you're correct, Whitney, then we've been moving objects through time as well as space. Rather large amounts of time, too."

"Which sets up a far more complex problem than we initially thought," Yariko said.

Frank hitched himself up straighter, wiped his sweaty face on his sleeve, and reached for the VHF. He fiddled with the knobs and startled the others with a loud sound of static. "I don't know how we got here or what you people were up to," he said, and the others all looked over at him. "But one man is dead, one injured, and I can't raise anyone on the VHF."

Dr. Shanker glanced up at Julian and shook his head.

Frank closed his eyes for a moment, and then opened them and went on, in a slightly stronger voice. "I'll keep trying. We'll contact a passing ship, if necessary, to get a message out. But the immediate thing is survival. We need to find water, and think about the possibility of being here over night."

Yariko watched him with the distant look that Julian was all too familiar with. Then she bent over her sand calculations again.

"I don't think our friend here quite appreciates the magnitude of the problem," Shanker said in a low voice.

Frank reached over and abruptly pulled the twig out of Yariko's hand, to her obvious amazement. "If you keep drawing lines in the sand, we'll have a real magnitude problem," he said. "The time to start looking for water is before you're good and thirsty."

"I need to know where we are first," Dr. Shanker said. "Don't worry, we'll take care of you." He turned back to Yariko. "Continue, please."

Frank chuckled, but to Julian it sounded more derisive than humorous. "I rather think it'll be the other way around, from what I've seen so far. One dead, one injured, and you people drawing pictures while time goes by. Now look." He pushed himself up more with his hands. His voice took on a new sharpness. "I'm no physicist but I do know about wilderness survival and setting up camp. Ex-marine, you know, special forces unit. You," and he jabbed a finger at Julian and Yariko, "you can go find some water, and something to carry it in. Use your imagination. You," and he turned to Shanker, "might start exploring. Stay within easy hail. Look for any signs of people."

"Look here yourself," Dr. Shanker broke in, in an equally sharp voice. "Who put you in charge? And as for your radio, it may be a useless piece of plastic at this point. That's what we're trying to figure out. Don't you understand what just happened?"

Frank's face went bright red. "I understand your lab exploded and killed Ron, and sent us somewhere else. You probably didn't care who got hurt. Now you're more interested in those equations than in your own survival."

"Frank's right," Yariko said, quietly. She stood, brushing the sand from her hands. "It's our fault. But it wasn't intentional." She looked directly at Frank as she spoke. "I think there was a precise moment in time when this—result was possible, and we unknowingly hit it. Of course I want to figure it all out. But Frank's hurt and we're in some unknown place—"

"Or time," Julian finished for her. He sat down on a rock, pushing aside some branches. His head was pounding; perhaps from the explosion, or the brilliant sun and heat, or maybe from the overpowering scents around him: a low-tide sulfur smell from the beach, rotting vegetation from the woods behind him, and the sharp smell of the laurels and some unknown creeper that had been crushed underfoot. Feeling a tickling sensation on the back of his hand, he looked down to see a beetle crawling across it: a large, red and gold beetle with exceptionally long antennae.

He stared at it for an instant, and then held up his hand for the others to see. "It's one of your extinct beetles," he said.

Nobody spoke.

Julian continued his thought. "We've moved back in time. Way back, to when those ammonite species lived, and this beetle." He took a deep breath and looked at Yariko. "Tell me I'm crazy," he said, and it was almost a plea.

"You're crazy," Frank said. "But that's not my problem right now. We need to make a camp so we can get through the night." He reached up to a low branch of laurel and pulled himself to his feet, swaying. He gingerly tested his weight on the injured leg, grimaced, and lifted it again. "I need a crutch," he said. "Get me a good stick, will you? If you three won't get moving, I will."

Yariko rushed over to his side. "Sit down, please. You'll only make it worse. You can't walk yet."

"I'll check out the water," Julian said. "For all we know, it's a freshwater lake." He walked away onto the beach, slowly.

A few birds skimmed across the brilliant water, too far off to see clearly. The mysterious crocodile, if that's what it was, had not reappeared. Julian both dreaded and anticipated his first sight of the inhabitants. Half of his mind was still looking out for a rescue party, listening for static on Frank's VHF, wanting to just sit in the shade and wait for it all to go away again. But another part of him knew what he had seen in the sand and knew there was only one possibility. He thought of the fossils—not the tiny innocent shells, but the monsters of the Late Cretaceous—and tried to reject the idea that they were about to come to life. "Translocation," Dr. Shanker said. But how did time travel come in?

He stopped halfway down the beach and stood gazing at the water, his hands in his pockets, momentarily forgetting about his mission. Gravitons, temporal translocation, beetles; he knew these were important things to discuss. Humans have a need to know why things happen. After that, one can start asking what should be done about it; because humans also like to dictate their own futures.

Frank was ready to simply live, survive, regardless of where or why. Julian however was an observer. He had always followed his future rather than dictating it. Now, he needed to know where they were before he could be concerned with why, or what anyone expected him to do about it. He needed more proof.

"Whitney!" Dr. Shanker's voice boomed across the beach. Julian turned back toward the trees, annoyed. But Dr. Shanker wasn't just calling him back to talk. "Whitney! Get back here! Run!"

Julian stumbled toward the trees; as he entered their shadow Yariko and Dr. Shanker grabbed his arms and pulled him in. Both of them looked terrified.

"What?" he mumbled, uncomprehending. Then a sound from the left made him whip around in terror.

It was a bellowing roar such as only a huge and fearsome beast would make.

It was answered by another bellow, away to the right.

Along the beach from both directions, parallel to the water, came the crocodiles. The one on the left appeared first. It was at least forty feet long; later Julian guessed its weight to be about ten tons, which explained the vibrating sand under his feet.

They ran straight at each other. This was not the belly crawl of lazy animals sliding into the water. This was the true crocodilian gallop: their whole, monstrous bodies were off the ground, their legs a blur beneath them, their backs bunching and straightening while their thick tails snapped out behind; they were running like lithe cats, with an occasional bounding leap over a rock or a log. And while they ran, they bellowed. Modern crocodiles are among the most vocal of reptiles, and their Cretaceous ancestors were apparently no different: just larger. Julian put his hands over his ears.

Nearly in front of him, not forty feet away, the animals came together. Julian expected to hear the crack of a bone-shattering collision, but at the last instant they swerved in perfect tandem, and shoulder to shoulder, like a pair of well-trained horses, they headed for the water.

The splash of their combined tonnage hid everything for moment. When the crocodiles could be seen again, the larger one was on top of the smaller one, whose tail was curling up into the air; an instant later, they sank together. A cloud of bubbles rose to the surface, and then the water was still.

The beach was absolutely silent.

A long, long moment went by, that seemed like forever. Julian took his hands off his ears and cautiously lifted his head.

"Oh, my. . . ." Yariko was standing at his shoulder. She had a look of shock on her face that made Julian's heart race even faster, if that were possible. She grabbed his arm; her fingers dug painfully into the muscle. "Were they as big as I thought?" she whispered. "How do they—I mean, I thought that they—do they lay eggs?"

"My God! The place will be crawling with them," Dr. Shanker said. "What a nightmare." His voice was shaking.

Frank said nothing. He stared at the water with a grim expression as he clung unsteadily to his branch.

Julian stood, trembling, not even feeling Yariko's grip. He took one fearful look at the water, but there was nothing to see. Then it dawned on him. He had just seen the mating of *Deinosuchus*.

They moved farther into the trees.

"Deinosuchus—does everyone believe me now? We're in the Cretaceous, all right," Julian said, tripping on a fallen branch as he looked back over his shoulder at the beach.

Yariko and Shanker eased the gasping Frank back onto the ground. "If those monsters try to come in here, they'll find a bullet," he said, clutching at his gun.

Dr. Shanker snorted. "They won't even feel your bullets," he said.

Crocodiles could be fast on a smooth, sloping beach, but they could not easily clamber through terrestrial undergrowth, whatever their size, as Julian explained. Still, it was hard to feel safe, and they were all tensed, waiting for another deafening bellow.

"I've seen them in zoos, but. . . ." Frank looked white around the eyes. "It can't be possible. Nothing can be that big." He laughed shakily. "Well. One more thing to consider."

"The fossil skull found in Montana was six feet long," Julian said. "That corresponds to a fifty-foot animal." He was as frightened as any of them; perhaps more, since he had actually been standing on the beach, near where the crocodiles met each other, seconds before their appearance. But at the same time he felt the numbness lift. He felt alive, alert, with his observational skills honed and ready. This was like one of those dangerous dig sites where everything, even the safety of the team, depended on his skills. He was ready, now, to be a member of this team, to contribute to their survival, to lend his expertise however he could. He was excited.

Dr. Shanker cleared his throat. "Somehow, a monstrous version of something we've all seen, something modern, seems worse than the unknown. They must grow fast so they can keep up with the big dinosaurs." He added that lightly, as if dinosaurs were a joke.

"Dinosaurs do grow quickly, like mammals," Julian responded, taking him quite seriously. "They probably reach full size early in their lives, maybe in the first ten years. Crocodiles are reptiles— they grow throughout their lives, at the same rate each year. To reach forty or fifty feet, they have to be decades old; fifty or sixty years, I'd say."

"But these are much bigger than anything I've heard of," Yariko said. "How do you know they grow at the same rate?"

"They lay down annual layers of calcification in their bones, and these can be counted, much like tree rings," Julian explained. "Based on fossil bones, Deinosuchus, and Sarcosuchus, its cousin from Africa, grew less than one foot per year. Deinosuchus lived in western North America."

"Um, I hate to ask," Yariko said in a small voice, "but what other huge creatures might be roaming around?"

"You don't want to know," Julian said.

Suddenly Frank sat straight, staring into the bushes. He reached for the gun in his holster, which was lying beside him.

"What is it?" Yariko asked, staring all around her with nervous eyes.

"Something's moving in there. Not a person." Frank cautiously slid the gun out and lifted it. At the same time they heard the sound of crashing underbrush.

They jumped to their feet, except Frank who calmly leveled his gun at the noise. The animal was very close. Then the bushes parted.

"Wait!" Dr. Shanker shouted. "Don't shoot! It's Hilda!"

Hilda scampered out of the dimness of the forest, her tail between her legs and her fur clotted with mud. She was too frightened even to bark. She ran to Dr. Shanker and buried her head against him, shivering and whimpering as if she had done something punishable. Dr. Shanker crooned over her, smoothing her fur and working out the clots of mud. "Thought you were dead, you dumb old mutt."

"Look! She's been wading through mud," Frank said.

They looked at him without understanding.

"Streams," he said. "Fresh water. Come on, people. Enough talk, more doing. We should take a look around. We need water badly."

Julian looked at Yariko; her face was sweaty and sandy, and her lips looked painfully dry. About as dry as his felt. It seemed like days since his last cup of coffee early that morning; he'd had nothing to drink since. Making a sudden decision, he stood up.

"Don't go alone," Dr. Shanker said. "Yorko, go with him. Take the gun."

"Do you know how to use it?" Frank asked. He made no move to hand it to them.

"No." Julian looked at Yariko, who shrugged.

"Then you'll be better off without it." Frank leaned back and closed his eyes; but he opened them again immediately. "What are you planning to carry water in?"

Dr. Shanker cleared his throat. "Ideas, anyone?"

Julian looked down and pushed some dirt around with his foot. He was suddenly very, very thirsty. He didn't want to have to think of how or where to find water. He just wanted some.

"Shoes," he said, suddenly. Yariko made a face. "Unless someone comes up with a better idea. . . ."

Dr. Shanker pulled off his loafers. "I'll drink out of my own, thank you," he said, handing them to Yariko. Julian took Frank's, feeling slightly foolish at his own idea, and he and Yariko set out.

"Frank was right, wasn't he," Julian commented in a low voice. "We waited until we really needed it to begin searching for water."

"He's very tough," Yariko said. "Knows more than you'd think for his age; I've talked with him in the lab a few times. Good thing he's with us. Good for us, I mean," she added hastily.

The light was dulled by the crowded trees and shrubs. It didn't feel like a bright afternoon in the forest.

"I don't like the gloom," Yariko said, her eyes darting from side to side in the low woods. "I feel like something's about to jump out at me."

"Spoken like a primate," Julian said. "A creature of the daylight. But we're much safer in the dark, out of harm's way. Witness the crocodiles; they're active in the heat of the day."

But he had to agree with her that the woods looked frightening. The trees were stunted and bent, almost bushes, the trunks branching close to the ground and lifting up a crown of leaves just above their heads. An intense gloom gathered beneath the thickest parts of the canopy. Here and there nettles or bushy scrub scratched at their faces, or forced them to turn aside altogether to find an opening. They passed a few large conifers, junipers maybe, twisted into

48

bizarre shapes, like gnarled old people. The ground was sprinkled with dead leaves that made it impossible to walk quietly. The cheerful patches of buttercups, growing near the edge of the forest, were absent in the gloom. It was no comfort that the smells were those of any ordinary forest, perhaps with a stronger overtone of decay; in fact, that made the whole situation seem more bizarre.

Julian's foot came down with a splash, and he jumped back. "Water." It was a brown puddle. He stooped and skimmed a little from the surface with his hand, and then sipped it.

Yariko hung back. "How is it?"

"Muddy," Julian said, making a face. "But it's not salty. Let's go a little farther and look for a clear pool."

The ground became soggy, squelching at each step. Their sneakers and socks were quickly waterlogged. Ferns grew thick everywhere, giving off a pungent odor when crushed. After a few hundred yards a low mud bank appeared: the edge of a pool. The trees hung out over the water, and on the farther side, about twenty feet away, the ground rose up again and the forest continued. The dim greenish light softened the far edges and blended the ferns and leaves on the opposite bank, giving a kind of opalescent sheen to the water. Tiny ripples and rings spread out on the surface where insects skittered about.

"It's beautiful," Yariko said. "But what about crocodiles? Could those big ones get up here?"

"No, they're way too big. There might be smaller ones though; in fact, some crocs stick to fresh or brackish water: the truly marine crocodiles would be dying out by now . . . if it's when I think it is, that is."

But there was nothing moving in the water. Julian knelt at the edge, his knees sinking into the spongy ground, and again scooped up a handful of water. It tasted clean, a little warm but wonderful after a hot day without a drink. He suddenly realized how thirsty he was. Yariko joined him and for a while they scooped water into their mouths, letting it run down their chins and dribble back into the pool, not caring if all the crocodiles in the world were watching. Then they sat back, panting a little.

Julian felt like a dry sponge that had just come to life in water. Now that his thirst was quenched, he was busy looking at the innumerable footprints in the muddy bank. They were jumbled and confused but dominated by the tiny, four-toed prints of a four-footed animal: clearly a mammal, perhaps the size of a squirrel. Other than that he could see nothing clearly; and nothing at all to indicate the time period. It might have been a twentieth-century stream on a subtropical island.

Suddenly Yariko started to back away, still crouching. "What's that?" she hissed

On the opposite bank, in the shadows of the overhanging branches, a pair of eyes gleamed red, without blinking.

"It shouldn't attack if we leave it alone," Julian whispered back. "It doesn't look very large."

They remained quiet for several minutes, trying not to startle whatever it was. Now and then the eyes shifted sideways a little, as if the animal were moving its head to watch from a slightly different perspective. Then the eyes winked out, and they saw the gray shape of the animal's body turning and flitting into the forest, making hardly a sound, only a papery stirring of the dead leaves. It was not reptilian. It seemed to be the size of a house cat, and Julian thought he saw a fluffy tail.

"What was it?" Yariko asked, still whispering.

"A mammal. But I don't know what kind." Julian leaned forward and filled Frank's rather large shoes with water. "This will be a fantastic place to explore. Tomorrow we should come back here and look for the footprints. If it really is the Cretaceous, think of how much we could learn. Think of the gaps in mammalian evolution that we could fill. And the plants. . . ." he gestured around them, ready to point out key species.

Yariko looked at him. "If we are in the Cretaceous, Julian," she said, "the rest of evolution hasn't happened. And when it does, we probably won't see it."

Julian felt his new excitement slip away, leaving a void as big as his body. Then the dread flowed back in. Something cold touched his fingers, and he started; looking down, he saw the water in the shoe break its surface tension and tilt a drop onto his shaking hand.

Yariko stood. "It's okay," she said. "We can look for footprints. There's no reason not to." She held out her hand to help him rise, muttering to herself as she did so, "It's not like there'll be anything else to do if we can't get home."

And it came to me that time itself was suspect!
—*Albert Einstein*

– Five –

1 September
2.31 PM Local Time

In one respect, it had been a very productive afternoon at the police station. University officials had cordoned off the basement floor but reopened the remainder of the physics building. They'd stationed a security guard in the lobby but agreed to let the local police team do its work. Since then, interviews with the other physics faculty members and with Miyakara's graduate student, an underfed-looking kid with wild hair, had clarified the cause of the initial explosion.

"Vibrations, probably sound, getting into the vault as it was powered down," Mark Reng had said. "That's why the door has to be sealed. But if they were inside making adjustments, they may have left the door cracked. . . ." Such explosions were unusual, unlikely, but perfectly possible.

But Sharon Earles was not reassured. Exploding equipment was one thing; disappearing people was quite another. There was still no word, sight, or evidence of those missing from the lab, and the afternoon was half over.

"A crime ring?" Hann suggested, jokingly. "They destroyed the lab to hide the theft?"

Earles checked each one off on her fingers. "Two physicists, a paleontologist, a security guard, and a German Shepherd. That's some crime ring. Maybe the dog did the actual theft so they wouldn't leave fingerprints?" She tapped her fingers impatiently on her desk. "And what about the missing half of the dead man? Charlie—" Earles paused, and then went on. "I know you're thinking about your brother Frank. Just keep in mind that he wasn't the one killed in the explosion. We'll figure this out."

Hann pulled over the cracked plastic chair and sat down. His face was somber. "It isn't like Frank to leave when he's on duty. He wouldn't. He just. . . . He'll show up. Tomorrow at the latest."

The boy was taking Frank's mysterious disappearance OK after all, Earles decided. But that was only because he wasn't the imaginative sort.

"Excuse me." The office assistant, Anna, appeared in the doorway. "There's someone here to see you, Chief. University safety team, or something like that." Behind her, a dark man craned his neck to see into the office, his eyes darting about in curiosity. Anna stepped out of the way.

"Yes?" Earles said. "Don't stand in the doorway. If you have something to say, come in and say it."

The man stepped into the room and held out a card. "Miles Gudgeon, University OSHA representative," he said. "You're the officer doing this investigation?"

"University what?" Earles ignored the proffered business card.

"OSHA. Occupational Safety and Health Administration. We're run by the Department of Labor."

"Ah. I was expecting a visit from you," Earles said. "Please have a seat, Mr. Gudgeon." Strange name, she thought. Strange man for Creekbend. No one wore business suits in this town.

Hann hastily stood and offered up his dilapidated chair. The man declined it. "I don't want to stay. I just wanted to leave my card, and tell you I'll be doing a survey of the lab in question. The main office is sending a team out here. They'll be traveling tonight. You'll see them first thing in the morning." He tossed the card on the desk.

"Not so fast," Earles said. "I'm conducting an investigation of a death. Possibly of several deaths. Your pack of officials will wait until I've finished."

Miles Gudgeon sighed. "I'm sure we can work something out. I'll be in the physics building, doing my own investigation. Tell your chief to call me."

That'd be me, pal, Earles thought, but she didn't say it. The little man wasn't worth the effort. However, he had the authority to do his own survey, she knew, and OSHA, despite the silly acronym, had the power to take the whole thing out of her hands. Well, she had one day and a night to do her own work. She didn't trust this Gudgeon with his little mustache and business suit.

"Sergeant Hann will accompany you," she said, ignoring Hann's uneasy look. "He'll see that you touch nothing. You may take notes but cameras and other recording devices are not permitted at the scene yet."

Before the man could object Earles ushered both him and Hann into the hallway. "Keep an eye on him," she said to Hann.

· · · · ·

The sun was well in the west when Yariko and Julian returned with their dripping shoes. A camp of sorts had been staked out by Dr. Shanker: a half-ring of large stones that Julian knew he himself could never have lifted. Hilda lay in the center chewing on a stick, looking as comfortable as if she were in her own front yard. Shanker and Frank accepted their water-filled shoes and drank gratefully; they didn't even make faces at the taste. Hilda got a loafer-full of her own, which she lapped noisily without bothering to stand up.

"We've gathered enough brush for a fire," Dr. Shanker said, indicating a pile of twigs and sticks. "Assuming we can even start a fire, of course. Frank's been trying to raise civilization again on his radio; needless to say, there's been no response. And I explored the beach a bit. There are no signs of human life here, not even distant noises. Incidentally, I've been unable to find any scraps of metal from the lab, so we may be reverting to the stone age."

"Stone age?" Julian laughed, and the sound was slightly hysterical in his own ears. He stopped abruptly. "Assuming we were sent somewhere by your experiment, what might be our chances of getting back?"

"That depends on where we are," Yariko said, sitting on the ground beside Frank. "And if we're really in a different time. Those crocodiles have me nearly convinced, but still. . . ."

"Giant crocodiles, birds with teeth—oh yes, I saw they had teeth—muddy water: it's beginning to look like we came to the wrong beach for vacation." Dr. Shanker waved his hand in a dismissive gesture, as if this was all a trivial annoyance, and sat down to put his loafers on. The swelling on his face was getting worse, but he'd scrubbed most of the blood away with a handful of the not-too clean-water, and he looked quite a bit better.

"But we don't want to jump to conclusions too soon," he went on. "Whitney may be convinced, but then this time period is his specialty. I say we figure things out now. All the more, indeed, now that we've met the neighbors." He turned to Frank, who was looking much better for the drink of water. "Am I allowed to talk physics now? Do we have your permission to discuss how we got here, and how we might get home?"

Frank shrugged. "Certainly, we need to be active in our own cause. Can't count on a rescue team showing up for some time, if at all. But this ring of stones is hardly a shelter; and I don't know about you, but I'm getting hungry. Finding food may not be so easy."

Julian's stomach gave a loud growl. "Sorry," he muttered, embarrassed. "Look, we've been here a few hours already, and I have a lot of questions. I want to know if this—translocation, is that what you called it?—if this is theoretically possible. And if so I want to know how we get home. Also, we should explore more, look for key species to support my time estimate."

"Typical scientist: wants to know everything," Dr. Shanker said. "We've only just realized that we were actually grabbing beetles and such from another location, and reassembling them, as it were, in the vault. Now I don't know what to think. Yorko: could we have temporal rather than spatial translocation going on?"

Yariko looked blank for an instant, and Julian could understand why: the mental transition from *Deinosuchus* and finding water to the far-off lab was a hard one. She was seated on a fallen log, looking tense but not panicked. She let out a big breath and then focused on explaining the experimental results. She spoke plainly, bringing home the situation in a way that was all the more shocking for being plain.

"Spatial and temporal translocation occurring together has not even been hypothesized. But whether we've moved through space or time, or both. . . ," she glanced at Julian, "we may be able to get back to Creekbend. But the odds aren't good." She paused, looking around at each of the others.

Frank grunted and then continued to look around him with darting eyes as if he was guarding an outpost under threat of imminent attack.

"Go on," Julian urged. "Tell us the details. Could you have predicted this?"

Yariko shook her head. "Of course not. It's all so new. But we do know a few things that might help." She flicked an insect off her leg and went on. "As you know, we've been producing strange samples in the graviton vault. I've been studying these objects for a month now, and . . . they don't persist. They disappear. I would put them in a bottle on the shelf and then come back the next day, and the bottle would be empty. My own guess now is that the objects were reverting to their original spatial coordinates—the place they came from, that is. If that's the case, and such translocated objects aren't stable, then perhaps we aren't stable in this place either."

The others were silent a moment as they took in this idea. Frank stared at her, looking at the same time hopeful and resentful, angry and interested. Julian hardly knew what he felt himself; he wanted a full explanation, and he wanted to become unstable and "revert" as it seemed they might; but he was also beginning to feel a mounting excitement that made breathing difficult. If they had really fallen into the Cretaceous world, he could learn more in an afternoon than in a lifetime back home.

Yariko took another deep breath and continued. "Even if my hypothesis is correct, even if these objects were returning to their original place, we are still far from safe. Only one-third of the samples ever reverted. That means that of the four of us, only one or two might possibly revert, or none of us. In addition, the equipment was in perfect working order when we produced the samples in the vault. We don't know what condition it's in now, after the explosion. So even if we could revert, the vault may not be there to take us."

Julian's excitement dissipated as quickly as it had come. "You mean, if the lab was damaged, we might be stuck here forever?"

There was a sharp clicking sound and they all turned, startled, to look at Frank. He was turned slightly away, as if he was shielding something with his body; the VHF antenna stuck up over his shoulder. Noticing the sudden silence he looked around, the radio clutched in one hand while his other hand fiddled with the knobs.

"Oh please," Shanker began, but Yariko stopped him.

"It doesn't do any harm to try," she said. "Let him. He knows how to use it, which is more than we do."

Julian thought he understood the man's nervousness; he himself was playing with his pocketknife to keep his hands busy.

"We'll obviously spend the night here," Dr. Shanker said grimly. "More than one, I should think." He reached out and pulled on Hilda's ear, eliciting a thump of her tail.

"But you say the beetles and such were unstable," Julian said. "They went back to their original location, or time, or so it seems. Then we will too, right? That beetle you called me about," and his voice got stronger as he remembered this fact, "the one I saw the picture of—it reverted in a few minutes, didn't it? It was gone by the time I got to the lab."

"There's more to it than that," Yariko continued. "I'll draw you a diagram so you'll understand the chances a little better." She snatched a twig from the ground and began to draw rapidly in the soil. Her hand shook slightly, and she stopped and clenched her fist for an instant before continuing. That slight tremble, almost controlled, woke Julian up nearly as much as the crocodiles had. He realized that Yariko was afraid too.

But when she spoke her voice was the same as always. "The objects first appeared in the vault; this square here. At first I kept them in jars in the back room—here." She drew a quick sketch of the lab's layout. "Most of them disappeared within minutes, hours, or days. After the first few days I began to study the patterns of disappearance. In fact, I spent nearly a week on this, at ONR's unknowing expense. Probably killed my career already." Yariko shook her head and Dr. Shaker gave a bark of laughter.

"So, what did you find out?" Julian was beginning to understand, but he wanted her to get to the point.

Yariko continued. "First, if we leave an object at its initial location, i.e., the vault, the reversion never occurs. It needs to be a certain distance away, and that distance depends on the mass of the object. Bigger things have to be farther away from the vault.

Second, the bigger something is the longer it persists. That's why we still had that rock to show you, Julian. It was the heaviest thing we'd produced, and it didn't revert."

"What she means," Dr. Shanker said, interrupting, "is that in order for us to have any hope of 'reverting' back to the vault, we'll have to calculate the right location, and we'll have to get ourselves there at the right time, based on our mass."

Julian frowned. "And naturally you need your IBMs for this sort of calculation?"

"Not at all," Dr. Shanker said, scratching his beard with a raspy sound. "The tensor equations are elegant enough . . . brilliant . . . it's a terrible shame that they don't give Nobel Prizes posthumously, or I would surely get one. Yorko, too," he added hastily.

Yariko nodded absently and went on. "The temporal latency is easy to calculate, because it increases predictably with the mass. For example, grains of dust disappeared within a few seconds. A pebble, ten grams maybe, about a third of an ounce, lasted forty-five minutes. If the same trend holds, then objects of our own size should persist for about two months. There is a certain margin of error, of course, about 5 percent, based on the size range we've had to observe."

"So . . . we wait for two months and see if we revert?" Julian thought about two months with *Deinosuchus* for a neighbor. A terrifying thought, but then again there was so much he could learn. . . . Another, unbidden thought came to him: two months with Yariko, nearly alone, no fiancé. . . .

"I wish it were that simple," Dr. Shanker said. "But there is the distance factor." He looked at Yariko. "Yorko's worked it all out."

"I did," she said. "The spatial calculation is more theoretical. Reversion only occurred if the samples were moved the correct distance from the point of origin in the lab. The direction was crucial: the objects had to be displaced along a narrow band that was angled between 2 and 4 degrees north of west. It took me a while to figure out why the direction was so specific, but the reason is quite simple. Months back, when I first built the equipment, I aligned it against the back wall of the lab, which happened to face exactly two degrees north of west; and by chance, I'd stored the very first samples along that back wall.

"But the distance—there the computations become more difficult.

The equations are hairy—messy—very complicated. There's no precedent for temporal translocation. Even now I don't know how it would figure into the equation. If we assume it works the same as simple spatial translocation, we can calculate, based on our weight, how far we have to displace ourselves."

She paused and looked around at the others, but she almost appeared not to see them. Strangely, her abstraction comforted Julian: this was the Yariko he was most familiar with, absorbed in complex thought processes and equations, as excited about calculations as he was about new fossils.

"How does that work?" he asked. "Why does it depend on mass?"

"I just want to know what we need to do to get back to South Dakota," Frank broke in. "And how we plan to set ourselves up for the night. It's already late afternoon."

Julian ignored him. "Can you figure out how long we'll take to revert, and where we'd need to be, from our average weight?"

"Yes, I could," Yariko said, simply, as if it were nothing to work out such things with a twig in the sand. "Give me half an hour without questions and I might have an estimate. But—"

"Good," Dr. Shanker said. "I'd work on it with you, but you get annoyed when I bother you." He turned away from her. "Whitney, let's put together a camp. We're in the boy scouts again, eh?"

Frank chuckled.

Julian had never been in the boy scouts, and he wasn't too excited about setting up a camp. Now that he was cooler and less thirsty, and Yariko was working out a plan for them to get home, he felt a great desire to explore this fantastic place. How could he possibly leave without learning everything he could first? Why, the knowledge he'd be able to take back. . . .

"Whitney! Stop daydreaming and help me," Dr. Shanker barked.

The sun was reaching down to the forest and the light in the underbrush was rapidly getting dimmer. Frank couldn't help much, but he snapped out advice as he tried to strike sparks between his pocketknife and a rough stone. Julian noticed that he used the file rather than the knife blade. A small pile of dry brown leaves and twigs was ready to take the spark.

"Look at this," Julian cried suddenly as he turned over a large rock. He stood and turned to Dr. Shanker. "There has been somebody here. Look."

On a flat part of the rock lines had been scored. They were partly filled in with soil but still plain to see: straight and slanted lines like text on a tablet. Somehow, Julian felt sure it was a word, a sign, an indication of human presence; but he couldn't make sense of the marks. He squinted at the strange lines, his mind unconsciously adding a few strokes to turn them into English letters.

"There's an A," he said, stooping and tracing it with his finger. "And here's another one. This here could be a Y."

"Whitney!" Dr. Shanker's voice was sharp. "You're as bad as Frank. Stop trying to read the rocks and help me get a wall built. Seeing letters in a rock. . . ." With some difficulty he lifted the stone and carried it, bowlegged and staggering, to their chosen campsite.

"But I saw. . . ." Julian followed, trying to protest.

Shanker dropped the rock and stood breathing heavily and shaking out his arms. "You of all people ought to know how easy it is to see false meaning in natural objects. Happens all the time. I have a piece of driftwood at home that looks just like Ronald Reagan's head." He chuckled. "What would it mean, anyway? A rescue party came back for us, but accidentally got here before us, so they left us a message? Those scrapes on the rock aren't new. Not a very effective rescue party, if you ask me." He stumped off to find more rocks.

Julian stooped again and turned the rock over, with some trouble. On a second look, the lines weren't so straight or patterned. They were just gouges, natural mechanical weathering. He sighed, and followed Dr. Shanker. Perhaps he was looking for any excuse not to believe they really were stuck in the Cretaceous. He was careful not to look closely at rocks after that.

By the time Yariko looked up from her dusty equations, they had a low wall of stones and sticks, two feet high, which half enclosed a flat area big enough for them all to lie down in. Several thick bushes screened off the end. Frank was directing them in setting up tall stakes between the stones when Yariko threw down her twig and said, "All right, I have some numbers. But I warn you it's not promising."

She joined them in the small semicircle. "Using our average weight, I was able to calculate the distance that we'd need to travel. I did the calculations twice. I believe the distance will be two times ten to the fifth, or 200,000 times, that observed in the lab, which was eight meters or about twenty-four feet. In other words, 1600 kilometers. We need to displace ourselves 1600 kilometers in order to reach the correct location for temporal reversion."

"That's a thousand miles!" Julian said. "Are you out of your mind?"

Yariko gave him a lopsided smile. "The distance is the least of our problems. The difficulty is getting the right angle. Being half a degree off north or south when we set out could put us hundreds of miles off by the time we reach the end of our line. There is also the question of confidence in the equations. Putting it all together, there's really very little hope."

"How confident are you?" Julian knew she was the most critical judge of her own work, and he didn't want her ideas, which might be their only hope, dismissed because they didn't meet her own too-high standards.

"Well, the numbers are what they are," Yariko said, with no modesty. "There's no problem with the calculation. But I have reason to doubt the applicability of the formula I used—"

"Just give it to us in English," Frank said. "Are you saying you made a mistake?"

"Of course not," Yariko said, and Julian couldn't help grinning at her shocked look. "There's no reason for mistakes in such straightforward equations. There's a check on the estimate, too: if my calculations are correct, the location we need should be exactly where our original mystery samples came from in the first place."

"Ah!" Dr. Shanker looked up with sudden interest. "But of course."

"Why of course?" Julian looked from one to the other. It seemed almost too easy, too coincidental.

"Because the settings in the experimental run were always the same," Dr. Shanker explained. "We were obtaining objects from one place, the same place each time. In order for us to be obtained, as it were, and end up back in the vault, we have to be in that exact spot. Assuming the settings are still intact, of course."

"Of course," Yariko went on. "Almost too easy. Except for one thing." She paused and looked at the others. "I had the brilliant idea of sending samples to the geology department, for analysis.

60

Bits of rock and dirt from six samples that hadn't reverted, from independent experiments. I expected the material would have a uniform composition, and would be from the same origin. I was right. So was Julian, with his guess on that pebble: the samples were igneous; apparently of recent volcanic origin. Very recent, even."

Yariko suddenly closed her fist around the twig and crushed it. "Of course, there are no volcanoes within thousands of miles of Creekbend, in any direction," she said, and her voice went up a half-octave. "Certainly not active ones. So the thousand-mile distance estimate doesn't match up with the place our samples came from. My calculations, therefore, seem to be off. I've checked them over several times, but I can't see why the formula should be any different." She opened her fist and the bits of twig spattered in the dust.

Julian watched Frank's fingers picking at the ground beside him. They lifted a smooth gray pebble and dropped it again, over and over. Dr. Shanker began to mutter disparaging comments about the geology department. Hilda laid her head on her paws and sighed.

"There are volcanoes," Julian said.

Everyone looked at him. "Not in our own time, but in the Late Cretaceous. The mountains—the whole Cordillera—are the result of many geologic events. . . ."

"Never mind the geology lecture," Dr. Shanker said. "Cut to the point."

"Volcanic activity on the surface, a thousand miles west of here. . . . Yes, that would put you at the easternmost fringe of the Rocky Mountains. The Rockies are partly volcanic in origin. In fact, there's a formation in southwestern Montana called the Boulder Batholith. I've been there to core for samples. In a straight line, the Boulder Batholith is about a thousand miles west of Creekbend. Maybe nine hundred miles.

"That puts it almost exactly where you calculated. I can't see where else your samples could have come from, in this geologic time—unless your distance estimate is way off. In that case they could have come from Alaska, Baja, or anywhere in between."

Yariko smiled again. "We could have saved you the travel for your cores, and given you fresh samples right there on campus."

Dr. Shanker interrupted her. "Yes, anywhere. Precisely. We don't know if the distance estimate is correct. Even if it is, we don't know if any of us will revert. We don't know if the vault will be ready for us. It's a long hike for a very slim chance, and the landscape in between is bound to be dangerous."

"The landscape will be dangerous wherever we are," Frank said. "On the other hand, I'm not sure we need to move from here. Camping is safer than traveling, especially into the unknown. We could be anywhere. We don't know what it's like further inland."

Julian thought hard. "I know something of the terrain," he said, "and that will help us find the right spot. I think Yariko's right, and we're actually still in Creekbend. Does that make sense? Creekbend is—will be, I should say—in the northern part of the state. What are we, twenty miles north of Roscoe?"

He began to place small stones in the soil as landmarks. "West from here will take us into Montana, through swamps and rivers and streams; wetlands, I should think. Nothing like the Montana we're used to. The elevation, for one thing, is much lower. I would expect to get into more dry terrain after about six hundred miles, and finally rocky hills in the western part of Montana. Young versions of the Rockies, that is."

Julian looked up from his map, flushed and excited. His eyes fell on Frank's crudely splinted leg and he fell silent. There was an awkward pause.

"How big are these structures, these batholiths?" Dr. Shanker asked after a moment.

Julian looked away from Frank. "Massive, usually. They form the cores of many mountain ranges. But this one is relatively small, only about five hundred square kilometers."

"Small! That's a sizable target. It almost makes the journey seem possible. You've half convinced me. Will we recognize it when we get there? Will it give us fireworks? Eruptions? Smoke and lava?"

Julian shook his head. He thought he felt Frank's eyes on him, and he didn't dare look at the man. "Not necessarily. Most of the activity should be below the surface. We'll see low hills and exposed, rocky terrain. We'll have to keep a sharp lookout for volcanic debris . . . and we'll need to keep in a straight line." Then another thought came to him, and a sense of hopelessness with it. "Two months . . . how can the vault stay set up and running for two months? They'll do something with it. They'll never leave it untouched." He tossed away the last stone and folded his arms.

"Oh, but it won't be two months in real time," Yariko countered.

"I mean, in our time. I mean, in their time."

"What do you mean? Time is time," Julian said, not ready to hope yet.

"Like . . . a different dimension?" Frank asked, trying hard to follow.

"Well, sort of." Yariko thought for a moment. "This is somewhat unconnected to what I was studying, but it comes in handy. I used to work on time travel. Models, I mean. Theoretically, if you could go into another time—gravity has a complicated relationship with time, you know, so it may be possible with a graviton vault—you could calculate the parallel movement of days in the two different time periods."

Julian's head began to feel too full. "You're losing me," he said. "Give us the equations some other time. Just tell us what it all means."

"How about I give you the simplified version," Yariko said. "I want you to believe me because you understand it, not because I say so." She picked up another twig and smoothed the dry soil, scattering Julian's pebbles. "Think logarithmically," she said, and wrote a series of numbers: 1, 10, 100, 1000. . . . "Each is ten times the other, right? Those are years. We're at negative sixty-five million years, or 0.65 of one hundred million years. That means our time will move much, much faster than time at year zero—modern time, that is—when they're running in parallel. And we can calculate that difference."

"Wait. That isn't right." Now Julian's head was beginning to spin. "A year is a year, at any point in Earth's history. The planet didn't orbit the sun a million times faster in the past, and life wasn't speeded up. Time has always been the same."

"Of course it has," Yariko said. Frank shook his head and bent over his tools again, striking them together with a grating sound that hurt Julian's teeth. Yariko continued. "Time is time. But time travel is another thing entirely." She looked around at the others. "Think of it as a long straight rope that can be folded," she said. "Time travel requires folds so that linear time differences meet ever so briefly, allowing mass to be transferred. The time of your origin folds differently from positive or negative times around your origin—the future or the past, that is. The differences depend not on absolute dates but on the length of time you've traveled, and how long you stay there. The greater the time jump, the greater the difference in 'folding.'"

Yariko wiped out all the zeros and ones in the dirt and began to draw equations. "So, using months, and sixty-five million years, two months spent in this time would put us back . . . in about twenty hours modern time. That's rough of course; it could be eighteen hours, or twenty-two. I do need IBMs for this sort of calculation," she finished.

Julian tried to make his mind work again. "So . . . we might show up in the vault twenty hours after we disappeared? Approximately?"

"Approximately," Yariko agreed. "Certainly within twenty-four hours. The lab just might be the same after twenty hours."

There was a silence.

Frank finally succeeded in striking sparks, startling them all with the new crackling sound, and began patiently nursing a tiny fire. "The bottom line is this," he said, looking up from his carefully cupped hands. "Travel or stay, we have work to do right now. We can't count on being rescued immediately; and if you're serious about this thousand-mile walk, there's a lot you'll need to learn first."

Only Yariko was able to meet Frank's eye. "But, Frank, your leg. . . ." She trailed off into silence.

"It'll heal," Frank said. "I've had broken legs before. Anyway, I don't intend to slow you down. Nobody has to take care of me."

Julian fidgeted with his hand in his pocket. Would they have to leave Frank behind? No, that was unthinkable. But so was waiting around a month for his bones to knit. And if he tried to keep up with them, he'd only prevent the leg from healing.

His fingers in a front pocket came into contact with something smooth and round. "My compass!" he said. "My pocket compass. Look." He pulled it out. The glass face was cracked, but still intact.

Dr. Shanker jumped up and took it from him. The low sun, filtering through the foliage, reflected off the glass. "A compass. That's wonderful. Whitney, you may be of use yet."

"Let's not get too excited," Yariko said. "Remember, the whole thing is quite untested. It's not remotely theoretical; it's barely hypothetical. And the lab may have been totally destroyed—in

which case it couldn't be repaired in twenty hours, even if someone tried."

Dr. Shanker interrupted. "Whitney. If that's east," and he pointed toward the water, "which it seems to be by the sun, then this
compass is pointing south. It must be broken." He handed the
compass back, looking disgusted.

Julian stared at it a minute and then grinned. "Magnetic rever-
sal," he said.

"What?"

"The earth's magnetic field flips every so often, so the poles
switch. This compass was designed for magnetic north being in the
Arctic, and magnetic south being at Antarctica. In this time, the
poles are reversed."

"Then. . . ." Yariko sat back down to think. "Then we really are
in a different time."

"Will it still work?" Dr. Shanker asked.

"Certainly. Only south means north." Julian looked at the compass and then pointed dramatically behind them, into the undergrowth. "Two degrees north of west: that way lies our road."

There was a rustle in the bushes and Hilda's head appeared.

"Hilda! Where have you been?" Dr. Shanker jumped to his feet.

Julian hadn't noticed the dog leaving, but in any case she was
back, carrying a small bundle of bloody fur in her mouth. She
stopped abruptly as they all stared at her; then she backed away several feet, fortunately outside of their makeshift wall, and lay down
with one paw over her kill. Julian itched to take it from her and
see what it was; but the dog began gnawing on it, slobbering and
crunching in the most sickening way. Julian looked the other way;
Yariko's face was scrunched up in disgust.

Dr. Shanker cleared his throat. "Speaking of food. . . ," he said,
and stopped.

"I'm getting quite hungry, now you mention it," Yariko said.
"Not that I want something dead to chew on. What are we going
to do?"

Julian suddenly felt weak and empty. He hadn't eaten since early
morning.

"OK Frank, you and your gun are on," Dr. Shanker said; but
Frank shook his head.

"There are only four bullets," he said. "We might use them all trying to bring down one animal. Better think of another way to hunt."

They looked at each other helplessly. "Spears?" Dr. Shanker said. "Knives, anyone?"

"My father was a master bowmaker," Yariko said. "I could try, but they'd be crude—"

"And dinner wouldn't be until next month," Dr. Shanker broke in. "Whitney, you know about these animals. This thing Hilda's caught—could we catch one of those? It doesn't seem to be making her sick . . . yet."

Julian shrugged. "Mammals in this time are mostly nocturnal. I'm not sure we'll even see them before full dark. Hilda probably sniffed it out of some lair." As with the water earlier, he realized they should have brainstormed before they actually needed the food. "I saw vines," he said at last. "We could try making traps . . . fishing nets. . . ."

"What I wouldn't give for a good hunting rifle," Dr. Shanker began, when a strange sound, an angry squawk followed by a thud, made them start.

Frank had lobbed a stone at a large bird that was perched on a rock just outside the trees, on the edge of the sand. As they watched, the bird tumbled off the rock and flapped madly in the sand. Then it stopped moving.

"Well, somebody get it before a croc does," Frank said, taking out his pocketknife. "It's big enough for the four of us."

They dined on *Ichthyornis*, as Julian described it, grilled over their little fire of dead leaves and sticks. It was a fierce-looking bird, with a long jaw and jagged teeth like a crocodile's; but it was only a fisher bird, and couldn't have harmed them. It did not look appetizing when plucked and gutted; but Julian said he'd once eaten roasted shearwater in New Zealand, where it was a delicacy, and found it quite tasty despite it's being a fish eater. *Ichthyornis* however did not live up to its distant descendent: it was an oily and fishy meat, not very pleasant even to very hungry people.

The light was nearly gone by the time they finished the bird and cleaned their hands as best they could. All too soon the sun was disappearing behind the forest, and the darkness began to come around them.

"I can't decide if I'd feel safer on the open beach or under the trees," Yariko said as they looked around in the gloom. "But I suppose a huge animal would give us plenty of warning with all this undergrowth to crash through."

"It's not the big animals that we need to worry about at night," Julian said grimly. "But this spot is probably as safe as anywhere, unless we want to sleep in the trees."

"Tomorrow we'll need to be more prepared before dark," Frank said, looking up from feeding the fire. "We may be spending several nights out here."

"If you count sixty as several," Dr. Shanker muttered, but Frank didn't seem to hear him.

Julian looked at their two-foot-high wall of sticks and stones and shook his head; but when even Dr. Shanker admitted to the added psychological security, he realized it was better than nothing. "The only nocturnal hunters will be small mammals, so the wall might help," he assured the others, wondering silently if this were actually true. "The Deinos certainly won't be active at night; they're reptiles. The predators that could hurt the mammals—and us—are daytime hunters."

"I sure hope you're right," Dr. Shanker said. "Good thing we have an expert along on this expedition."

They decided on an order for keeping watch, and began their first night in the Cretaceous.

Despite his paleontological knowledge, Julian spent his two-hour watch huddled in fear, surrounded by an eerie cacophony of night sounds. Leaves rustled, branches snapped, trees creaked; and every few minutes, it seemed, a chorus of yips or hoots swelled all around him. He wondered if they were close to another water source, and if the mammals, and perhaps other things, things he shuddered to think about, were collecting there.

Frank's small fire was only bits of black sticks with an occasional red gleam, giving off no real light. Dr. Shanker gave a loud snore at irregular intervals, making Julian jump. The air was almost too warm but he shivered. He resisted the urge to wake Yariko, or even Hilda, just for the company. When he lay down after Frank took over he was more exhausted than he'd ever felt in his life. His brain was screaming for rest and every muscle ached. He curled up in the patch of earth they'd cleared of debris, and was instantly asleep.

From across the quad in Creekbend he saw *Deinosuchus* coming for him. It came out of the biology building and walked, elbows bent, along the sidewalk, students scattering in every direction. But no, it wasn't coming for him. It only wanted to get to the particle physics lab, so it could revert and go home. . . . But, he thought, how can it fit in the vault?

The crack of a gunshot brought Julian back to the Cretaceous.

Somehow, it has gotten into the popular culture that life was more abundant in the Cretaceous, back near the primal beginning of the world. This view is a holdover from the eighteenth century, before anyone knew how old the earth really was. The Cretaceous was actually a very recent period, and despite the hothouse climate and high sea level, was in many ways not that different from the Quaternary, the proper time of Homo sapiens.
—*Julian Whitney,* Lectures on Cretaceous Ecology

– SIX –

Julian tried to jump up but something was holding him down. Something heavy. He grabbed at it in fear and tried to shove it away.

He heard a gasp as of pain, right in his ear; then a hand pushed at his face. "Stop," he sputtered, trying to turn his head. "It's me. Get off me." Frank, maybe? He couldn't tell.

But it was Yariko's voice that came next. "I fell on you. Let me go."

Julian realized he was gripping her shoulders. He let go and sat up. "Are you all right? What just happened? Where are the others?" Even in his fear he felt embarrassment; Yariko had been lying right on top of him, and he'd grabbed at her . . . what would she think?

"All right, everyone calm down," came Dr. Shanker's voice from off to the right. "Hilda! Come back here and stay." He could be heard coming toward them. "Put that gun away," he went on. "What happened to the fire? Let's get some light."

"Frank!" Yariko called.

"I'm still here," Frank said. "Never far, you know. See if you can find any twigs and I'll make some light."

They scraped enough together for Frank to get a tiny blaze going. "It's best to collect wood in the daylight," Frank commented dryly as he carefully added small branches to the fire.

69

"Never mind that, what just happened?" Dr. Shanker came into the little circle of light. He sounded irate. "You nearly blew my head off with that thing."

"Actually, you were behind me," Frank said calmly. "Something attacked us, and I shot it."

Dr. Shanker snorted. "What's the point of shooting off a gun when you can't see a damn thing? I thought there were no predators at night anyway."

"There was something," Frank said. "It came out of the bushes at me, and it bit me. It wasn't Hilda," he added.

Hilda could be heard crunching on something off to the side. She's tough enough on sticks, Julian thought. Why didn't she give us any warning? "You got bitten? How bad is it?" he asked.

"Just a scratch. The thing jumped on me. It had claws, too. It was going for my face. You kicked out the fire right after I shot the thing," he added to Dr. Shanker.

It took some time to calm Shanker's annoyance and Frank's sense of mistreatment. He insisted he had shot the animal despite its invisibility; Dr. Shanker insisted that a small animal shot at such close range would have blown apart rather than run away.

"I hit something," Frank said. "I know the sound of a bullet hitting a body close by. Down one bullet, three left," he added.

Yariko took the next watch. There wasn't anything they could do about Frank's injuries until daylight, and he said they weren't bad. Julian settled down again, curled on his side, and lay a long time awake.

In the dim early morning the mystery was cleared up. The remains of Hilda's midnight snack, the front end of a vaguely raccoon-shaped mammal, had a neat bullet hole from one side to the other.

"That was some shot, by firelight," Dr. Shanker admitted. "You're either a crack shot or very lucky."

Frank didn't comment; he was polishing the gun.

He had a small but painful bite on his shoulder, a long gash on one cheek, and several more on his forearms. He sat very still as Yariko looked him over but he looked relieved, Julian thought, to be proved right.

"That raccoon thing was probably even more scared than we were," Yariko commented. "Imagine wandering around in familiar

70

Cretaceous territory and suddenly bumping into us." Even Frank had to chuckle at that.

He looked much better already, despite the night's adventure; but he could not walk. On his watch he had fashioned a very clever crutch for himself out of a forked branch with an added crosspiece, but he could hardly even hobble into the bushes to empty his bladder. Julian suspected the leg bone wasn't set properly.

"What's this?" Dr. Shanker said suddenly, stooping to lift something shiny from the dirt. "It looks like metal." He held it out.

Julian's hand went to his watch pocket. It was empty. He frantically slapped all his jeans pockets, knowing the compass wasn't in any of them, knowing that it lay in bits and pieces in Dr. Shanker's palm.

There was a hush as first Yariko, and then Dr. Shanker, realized what it was.

"Look for the magnet," Julian said. "Is it there?"

It was not in Dr. Shanker's hand, although Julian insisted on taking the pieces and looking through them for several minutes. He dropped to the ground and began feeling around in the leaves and soil. "We've got to find it," he said. "We can reconstruct it. The card's broken but we can make another . . . even a crude one, with just the Cardinal points. . . ."

But although he and Yariko both searched the area, sweeping the ground with Julian's pocket knife in hopes that the tiny magnet would stick to it, they found nothing.

They stood and looked at each other. Julian had an overwhelming feeling of failure, as if he'd done something terribly wrong.

Then Dr. Shanker laughed. "Who needs a compass? We'll navigate by the sun and stars. I know a few tricks for finding a rough latitude—and so do you, Whitney. I saw you yesterday, eyeing the sun's altitude. We'll be all right. I didn't trust that little gadget anyhow."

Julian realized he was right. The compass was convenient, but not decisive. Strangely, Frank was the only one who remained concerned about its loss; perhaps he was beginning to believe in their real situation. Certainly he hadn't turned on his radio yet that morning.

"How do you find the sun's altitude?" Frank asked. "That could tell us something of where we are."

Julian stretched an arm out toward the ocean horizon and closed one eye. "See my fist? I rest it right on the horizon—convenient that it's ocean—and the top, at my thumb joint, is roughly ten degrees above the horizon." He placed the other fist on top of the first. "Now I have twenty degrees. The sun is very roughly at twenty-two degrees, I'd say."

"What does that tell us other than the time of day?" Yariko asked, trying her own fists.

"It's a three-part equation, with one unknown," Julian explained. "Simple for a physicist. It's September second, right? So the sun is nearly halfway between its most northern point—our summer solstice—and its most southern point, which is our winter solstice.

"We've picked an easy date, anyway. We're clearly in the northern hemisphere and at noon today we'll see just how far north of the sun we are. Then we'll know our latitude, possibly within five degrees."

"We just moved sixty-five million years in time, and you think it's still September second?" Dr. Shanker snorted.

Julian shook his head impatiently. "It has to be," he said. "Or pretty close to it. I've been watching the sun. If we're anywhere near Creekbend, then the time is early September. Or would be if months were recorded around here. . . ."

"Whitney," Dr. Shanker said, "can the lecture. I want to know what we do now. Form a scouting party, before we begin the journey in earnest?"

"Breakfast," Frank said.

They looked at each other helplessly. Julian had a sinking feeling as he realized that each and every meal would pose the same problem. They couldn't expect to knock birds down with rocks every day.

But Frank already had the solution. "Tubers," he said, producing a fistful of pale roots that resembled over-large parsnips with feathery greens sprouting from their tops. "I pulled a few up during the night; they were digging into my back. They taste like a mix of carrot and radish. A bit tough, but that could be fixed by a little roasting." He began scraping together kindling for another fire.

"You ate them? Without knowing if they were safe?" Yariko sounded aghast.

"Certainly," Frank replied calmly. "After studying them a bit.

I've had to forage on strange plants before, you know. Roots are much less likely to contain poisons than leaves. Always go for the roots, if in doubt."

"Well, you're still alive, so we might as well dig in," Dr. Shanker said. "Whitney, how about a piece of your compass face? We could ignite a leaf with that bit of glass, now the sun's up."

The tubers when heated on a flat rock were almost sweet, very sticky, and quite filling. Julian's spirits rose again. This wasn't so bad after all. In fact, he couldn't remember having a more satisfying breakfast in a long time. He carefully did not mention the craving for coffee, and nobody else seemed willing to either.

They discussed the morning's exploration. Clearly, they needed to see what was immediately to the west, and make a plan for beginning their thousand-mile trek. Equally clear, though unspoken, was the fact that Frank wasn't going anywhere for a while, and nobody was sure what this might mean for their chances. Yariko voiced the opinion that after they explored a bit Frank should direct them in work parties geared toward basic wilderness survival. Julian wondered if she was only trying to keep Frank's spirits up; there was after all no point in getting settled in their present location.

They drew straws, or rather twigs; Julian and Dr. Shanker were chosen as scouts. Yariko would remain with Frank, and he would keep the gun. Dr. Shanker wanted to take it, if only to keep Frank from shooting the returning scouts by mistake, but Frank wouldn't give it to him.

"Please, be careful," Yariko said, as the explorers turned to go. "Don't leave Frank and me all alone—I mean forever." But she looked at Julian rather than at Shanker.

Julian selected stout walking sticks to beat aside the brambles and the two set off along the edge of the trees, Hilda trotting at their heels. The sun blazed across the sea, sparkling over the waves. Julian saw a grim humor in exploring the Cretaceous landscape armed with knobby walking sticks; but he didn't laugh. Near the water they saw one set of five-fingered tracks, with the trail of a heavy body between them, but no crocodiles appeared.

Not far down the beach a stream emerged from the forest. They decided to follow it back to its source. It was easy to track, a trough of mud snaking through the twigs and dead leaves of the forest floor. It led directly west, into the woods.

They plunged into the cooler shade of the trees. There was no sound but the crunch of sticks underfoot and the hissing of the surf, rapidly fading as they walked deeper into the forest. A few flies buzzed about, attracted no doubt to the smell of sweat. By late morning the birds were silent, resting from the heat, and there was no sign of any large animal. Hilda nosed along behind them, panting.

Despite the popular myth, the Cretaceous flora was not dominated by giant ferns. In fact, ferns were quite rare. Julian had already noted that most of the trees were in the laurel family, although there was no laurel *per se.* He now saw magnolia, and a scattering of ginkgos. The fan-shaped leaves of the ginkgos fluttered in the slight cross breeze, and the ground was carpeted with yellow leaves from the previous year. Dried and partially decayed nutshells crunched and crackled underfoot, sending up a slightly rotten odor. A patch of nettles grabbed at their clothes and stung the backs of their hands. Already the air was too warm.

They had not gone far when Dr. Shanker stooped to look at something in the dead leaves. Then he stood quickly and pointed. "Whitney. Is that what I think it is?"

Julian stooped also, and lifted a whitish object. He held it out on his palm. It was a two-inch tooth, curved and tapered, blunt at the tip from wear. "What did you think it was?" he asked.

"I think it belonged to a carnivore," Dr. Shanker said. "Something must've died here."

Julian studied the tooth closely. "Carnivorous dinosaurs shed their teeth all throughout life and grew new ones—hence the jagged appearance of their dentition."

"Can your dental expertise tell you what animal it's from?"

This tooth was very distinctive, and Julian had recognized the genus immediately. "See the serrations?" he said. "They're exceptionally large and hooked, along the front and back edge; a structure unique to Troodon."

"Large or small?" Dr. Shanker asked.

"One of the smaller carnivores."

"Small!" Dr. Shanker looked over at Hilda. "I used to think her teeth were big. This animal must be five times her size, to have a tooth like that."

"It isn't. It's smaller than us, in fact." Julian pushed the leaves around a bit but saw nothing else interesting. "Probably just the size to hunt puny animals like us," he added. "They're thought to be intelligent, and to hunt in packs." He spoke lightly but he shuddered as he let the tooth fall.

Dr. Shanker only grunted and tightened his grip on the walking stick. They continued into the forest, peering nervously through the trees; but the silence remained unbroken. They tried hard to keep track of the sun and keep as straight a line as possible.

As the ground became wetter, the character of the plants changed. Ferns grew in patches, and moss clung to the stones and to the trunks of trees. A few close relatives of the sycamore appeared, and their spiky round catkins lay scattered over the ground. Lianas hung from the branches.

Here they saw their first forest vertebrate: a snake about a foot long and the width of a pencil, which quickly slithered away. Julian was not much worried about snake bites, because he knew poisonous snakes had not yet evolved. The tiny, black-and-white flies were of greater concern, swarming up from the mud and biting at the backs of their necks. They picked handfuls of ferns from the bank to swat them away.

"Hunting in packs—just like these damn flies," Dr. Shanker growled as he slapped his neck. Then he pointed to something. "Hey Whitney, there's a rock sitting on top of another one. Think it's from your rescue party?"

Julian scowled. He didn't like being the butt of sarcasm; it happened too often in his life.

The mud became a small stream. Hilda, walking happily in the cool water, suddenly froze with one muddy paw raised.

"Hilda! What is it?" In normal life Dr. Shanker hardly paid attention to Hilda's frequent pauses. Things of interest in the dog world rarely mattered to him. Now, however, he seemed to lean on her keen sense of animal life.

In response the hair on Hilda's back rose and a low rumble came from her chest.

"Do you see anything?" Dr. Shanker whispered.

"I see a fallen branch." Julian cautiously pointed.

"Hilda doesn't care about branches. She's not dumb." As he spoke Hilda slunk toward the opposite bank, her ears back and her head low. Dr. Shanker reached for her collar but she slipped past him and scrambled up the muddy slope.

"Don't follow her," Julian said, clutching at Shanker's sleeve, but he pulled away and sloshed after her.

"Hilda!" he said, in a half whisper. "Come back!"

She had already slunk into the thicket and disappeared, and he pushed aside the vines to peer after her. He was silent a moment, and Julian waited anxiously, ready to run for a tree. Then Dr. Shanker said in a normal tone, "Whitney, take a look. It's dead, whatever it is."

Julian climbed the bank beside him. In the darkened cavern beneath the trees something lay sprawling on the soil. There was very little smell of decay. Hilda circled, growling.

It might have been a large lizard, or a small dinosaur, and Julian held his breath and stepped closer for a better view. He prodded the thing with the walking stick and turned it over a few times. A cloud of flies went up with an angry buzz. The head was gone and the scrawny neck flopped about as the body was turned. The hind limbs were much longer than the forelimbs, well muscled, hinged like the legs of an ostrich; one leg had obviously been feasted on. But the most gruesome part was the belly. It had been torn open, as they saw when Julian flipped the animal over.

Dr. Shanker wrinkled his nose and stepped back as the pile of intestines was revealed, covered with soil and bits of leaves. "What a way to be killed," he said. "Why wasn't it eaten?"

"I'm not sure," Julian said, also stepping back. "It may not have died right away. It may have tried to keep up with its herd, and the predator gave up, or lost it." He decided it was a bipedal animal, though it may have gone down on all fours now and then. From the tip of the tail to the tip of the neck it must have measured five feet. "I can't tell the species. The teeth would have clarified everything, but without them. . . ."

"You mean the thing could've run with its intestines hanging out?" Dr. Shanker sounded horrified. "Well, let's move on. Some scavenger might come along, and I don't want to be here."

They turned away.

Finally they reached more open water: a sluggish, oily stream passing between high mud banks. Right at the bank stood three gray old conifer trees of staggering size. They were so close together that a small animal must have carried the seeds to that location and left them, a few hundred years earlier. Julian studied them a moment and decided they were in the cypress family, some close ancestor of *Taxodium distichum*. More to the point, they were the largest trees anywhere nearby.

Dr. Shanker looked up into the trees also, as if studying their height. He paced around them with his arms folded. "Taller than all those sycamores, anyway," he said.

"Real sycamores haven't evolved yet," Julian said. "These are in the same family, but—"

"Evolved be damned," Dr. Shanker interrupted. "I'm calling them sycamores. You're lighter," he added, before Julian could volunteer. "Ready?" He cupped his hands as a stirrup.

Julian opened his mouth to protest, and then shut it again. He looked up at the enormous trunk seeming to disappear into the sky, and shrugged. "Thanks," he said. "I'll be out of reach of Troodon, anyway. Keep a good lookout down here." Stepping into the proffered hand, he balanced against the shaggy bark and strained to reach the first large branch.

It was too high. He wobbled on his one foot, scrabbling at the bark. "Give me a boost," he cried, and with his gymnasium muscles Dr. Shanker launched him upward so violently that his brains were almost knocked out against the limb.

The climbing was easier from there up. Branches jutted out of the trunk at regular intervals like spokes, although they were angled downward so that Julian had to cling tightly to avoid sliding off and crashing to the forest floor. After a few minutes of careful and slow progress he came out through the tops of the "sycamores" and vines, hot, scratched, and dazzled by the sunlight.

Dr. Shanker was a distant dab of color all but hidden by the intervening branches. "What do you see?" he called up.

"Nothing yet." Julian continued to climb. He thought he was about fifty feet up when he stopped. The tree continued on perhaps twice as far as he'd climbed; but even so the breeze was already alarming. At every puff the tree seemed to sway, and Julian's stomach jumped.

He shifted his grip and found a comfortable place to stand. Then he looked around.

The view was fantastic. Looking east he saw the tumbled green masses of the sycamore forest spreading out like a shaggy rug beneath him. Here and there through gaps in the canopy he caught the glitter of water. Beyond the forest lay a dark green strip of the laurel-like trees and beyond that, perhaps half a mile away, lay the beach. He could see a shimmering line of white where the waves came in, and then the vast quiet blue of the sea.

Since Julian already suspected they were still in South Dakota, probably in Creekbend itself, it was not hard for him to decide what he was looking at. At a guess without waiting for the noon sun he had put their latitude around forty-five degrees north, and the panoramic view now confirmed his suspicions. At this geologic time a vast seaway covered the central part of the continent, stretching north to south, connecting the Arctic Ocean with the Gulf of Mexico. Minnesota, Iowa, Missouri, Arkansas, Louisiana—all were under the Niobrara Seaway. A few million years earlier, and the Dakotas would have been submerged as well; but the inland sea had been drying up for some time now.

Three days ago, or sixty-five million years into the future, however one looked at it, the land below him had been dry and rocky, cut by roads and sprawling cities, blanketed with smog, smelling of car exhaust, littered with the bones of extinct animals. Now Julian looked out over thick green forests, mist rising in the morning heat, a crystal sea lapping at the shore, a blue sky with hardly a cloud. He was so entranced that he clung there for ten minutes, staring, scuttling from one side of the tree to the other, taking in all the incredible unending wilderness.

"Hey up there! Anything to see?" Dr. Shanker's patience had given out.

"Trees. Lots of trees. And the beach, and the sea. It's spectacular," he shouted back.

"Whitney!" Dr. Shanker bellowed; Julian grinned as he imagined Shanker pacing in fury down below. "I don't care about the damn beach! Turn around and look west!"

Julian turned gingerly around, put his arms around the rough trunk, and eased himself over to a branch on the other side. A piece of bark went in his eye and he blinked furiously for a minute, afraid he would lose his grip and fall. When his eye cleared he raised his head to look and gave a shout at what he saw.

They were on an island.

To the west lay a strip of water separating them, by perhaps a mile, from what looked like the mainland. He strained to see what he could on the far shore, but he couldn't make out any detail. The land was green, thick with trees. There appeared to be a cleft in the shoreline that might correspond to the mouth of a large river, but he could not be sure.

For a moment his heart almost failed him, and he closed his eyes in denial. An island. How much worse could their luck have been? Here was an obstacle before they even began. A mile's width of open sea might as well be an impenetrable wall.

Finally Dr. Shanker's voice boomed up from below. "Whitney! Hey up there! Come down now! Don't hog it all!"

Julian sighed, and began to pick his way down again.

Dr. Shanker took the news calmly. "Well, let's go have a look at the other side," he said.

Continuing downstream they soon reached the western edge of the island. The stream emerged from the forest and disappeared into a soggy delta speckled with ferns and bushes and swarming with flies. Being less than eager to slog through the mud and the flies, especially in the staggering heat of direct sunlight, they stood in the shade at the edge of the forest and looked out.

The channel separating them from the mainland might have been more than a mile, but there was a clear view of the opposite shore. They were looking at the mouth of a large river, its brown waters emptying sluggishly into the sea.

"An island. A Goddamn tropical island." Dr. Shanker stood with shoulders slumped, leaning on his stick. "I was hoping you were wrong."

"We'll build a boat," Julian said, surprising himself with the simplicity of the words.

"A boat?" Dr. Shanker now looked incredulous. "With what— sticks? And what kind of monsters live in that water?"

But Julian's mind was on the distant river. It came from the west, he was sure. From the higher, dryer ground where the Rocky Mountains were being born. And it led into the heart of a seething wilderness that no human had seen.

A low rumble, barely heard, interrupted his musings. Hilda appeared beside him and he realized the sound came from her chest.

"Whitney," Dr. Shanker said, casually. "We're not alone."

Julian turned his eyes from the shore and stared at the bushes behind him. Something was hiding there watching—a large animal, crouching in the mud.

He caught the glitter of sun from its eye.

Hilda crouched, her hair on end and her ears back, and then suddenly lunged toward the bush with a snarl.

At the same instant the creature burst from cover and ran. Julian was horrified by the size of this animal that had been hiding only ten feet away. It was larger than an African elephant. Its great tridactyl hind feet splashed through the marsh and sent a spray of mud in all directions, while it balanced by touching the ground now and then with its smaller forefeet.

The two men turned to run; but Julian stopped and clutched at Shanker's arm. "Wait," he said. "Let's not panic and get ourselves lost. It won't hurt us."

"Oh yeah?" Dr. Shanker sounded uncertain, but he paused and turned with Julian.

As the enormous creature lumbered off, Hilda barking her head off behind it, they could see plainly that it was an adult counterpart to the mangled body they'd found. It was a dull greenish-brown all over, blending perfectly with its surroundings. The back was marked with faint yellow vertical stripes, as if mimicking the dappling of sunlight in the forest, and the belly was plastered with mud.

For Julian, what gave away the type of animal was the shape of the head: the large eyes placed to the sides, giving it that frightened appearance of game, and the snout flattened into a wide bill. Suddenly it trumpeted as it ran, a great honking noise like a goose amplified ten times. Hilda came to a skidding halt, turned, and ran back to Dr. Shanker with her tail between her legs.

At the warning cry, the area all around burst into life. More of the creatures bolted from ferny patches, leaped from behind bushes, and even stood up in plain sight, where they should have been easy to see if it weren't for the marvelous camouflage. Julian and Shanker had been standing in the middle of a small herd. None were as large as the first one, and a few of them seemed to be quite small, Hilda's size, skittering through the mud, trying to keep pace with the adults as they splashed away and disappeared into the trees. Within thirty seconds the area was silent again.

"Hadrosaurs," Julian said in an awed voice, once his heart had slowed a bit. "Probably Edmontosaurus. There was no crest on the head." I've just seen my first live dinosaur, he thought, and had to suppress a hysterical giggle.

Dr. Shanker did not seem overawed. "Hm. Herbivores, I take it? What's a herd of such huge things doing on this tiny island?"

Julian thought for a while as they stood gazing across the water. "They probably swam over from the mainland to escape the large predators," he said at last. "Given the island's size and the dominance of Deinosuchus, it's unlikely any large carnivorous dinosaurs live here."

"Meaning they're all waiting for us over there," Dr. Shanker said, nodding across the stretch of water. He turned away, heading back, and Julian silently followed.

Most dinosaurs had four digits on the hand and three on the foot. For Troodon, *however, only two digits touched the ground, leaving a distinctive print. The third toe curved up, supporting a hooked blade, one of the creature's main offensive weapons. Although this animal averaged only about 50 kilograms, the size of a wolf, it was probably a highly effective and intelligent hunter. Judging by the size of the brain case, it was among the smartest of all dinosaurs. With its very large eye sockets,* Troodon *probably had keen vision in low light, enabling it to hunt at dawn and at dusk, and in the dim light of the forest.*
—Julian Whitney, Lectures on Cretaceous Ecology

– SEVEN –

It took Frank two days to make the journey to the western side of the island, moving slowly a short distance at a time with frequent pauses, and resting in the middle of the day. He remained alert and full of succinct advice; advice that they now knew was worth listening to. He seemed to know a great deal about survival in the wilderness.

It would have been better for the healing process if Frank had kept still for a week, but all had a common urge to get away from the beach and the giant crocodiles. They were also relieved to begin the thousand-mile journey, albeit a slow beginning, because by doing so they had made a decision, and felt their determination would keep them going. Frank was immensely cheered by the prospect of a river journey, at least at the outset. His leg might heal while they paddled upstream on a raft. Julian didn't point out that the river on the mainland might be very short, or make a turn and take them in quite the wrong direction.

Yariko and Dr. Shanker built a crude shelter on dry ground, far enough from the edmontosaur swamp to be safe from flies, with fallen branches and fern fronds. The group spent four days there, in the end, and everyone felt better "inside," although they knew the shelter would be meaningless to a predator. Frank kept both the gun and the VHF radio near him; he'd grasp them if there was a noise like an animal approaching. Otherwise, he sat next to the shelter making spears and various tools, or so he said. He'd get a small fire going to blacken the ends of straight sticks, which he'd then chop at with a stone, scattering charred wood everywhere.

Yariko became the water collector. She'd found an empty turtle shell about eighteen inches across that made an excellent pot for boiling water. With Julian's knowledge they managed a reasonable diet that included roasted mammals, a very few plant stems, and tubers. But it was Dr. Shanker who was the best cook among them: once he made a mammal stew, complete with crisp plant stems and a leafy spice, that was almost tasty, although not entirely free of fur.

Yariko was also the weapons master. She claimed to have watched her father at his bow-making hobby, and despite her colleague's derisive snorts, she determinedly collected sticks and vines and tried all sorts of arrangements, without any success. Julian would return to the camp and find the two of them sitting there, Frank surrounded by chips of stone or wood shavings, Yariko looking sadly at a tangle of braided vine and broken stick.

"This is hard work, this living like a caveman!" Yariko exclaimed on the second morning in their new camp, as she tried to chip wood with an axe of sorts created by Frank. Sweat was dripping off her upper lip and her T-shirt was soaked. Her hands were scratched and bleeding, her hair in its messy braid held bits of leaves and a twig, and dried mud flaked off her jeans every time she moved. Julian thought she'd never looked so beautiful, or so determined.

Yariko swung the makeshift axe again and the stone head flew off, fortunately missing Julian, and landed in a thicket. She finished the swing with the handle and then sat down on the branch she was trying to cut, and laughed.

"Stone age indeed. Do you know," she said, rubbing her arm across her forehead and leaving a streak of dirt in the sweat, "I've spent so much time lately inside a lab, living in my head with computers and calculations, it was almost like I didn't have a body anymore." She looked up at Julian with her face flushed and her eyes big and bright. "Here I feel so alive. Not just my brain, but all of me. My hands." She held her hands out, palms up, to show the calluses and splinters. "My arms. My legs. I can feel every muscle. I'd almost forgotten what my body was capable of," she finished, with another laugh.

Julian had never wanted so much to grab her in a hug . . . and perhaps more. He had certainly not forgotten what her body was capable of, when he used to watch her back at the university. There had been a distance between them even when Yariko was at her friendliest, as if she didn't entirely see him, or didn't let go of her mental calculations while talking to him. Or perhaps she was always thinking of someone else.

Looking down at her happy, very alive face, and smelling his own sharp sweat mixed with a whiff of hers, Julian suddenly wanted to stay in the Cretaceous forever.

And the Cretaceous was becoming more wonderful to him every day. At dawn and at dusk, he became an explorer. He sometimes brought in small animals to eat; but more often he brought reports of footprints or other signs of animal life around their camp.

Once he saw the distinctive footprints of *Troodon* sunk into the mud near the stream, not far from the camp. He studied them for some time, trying to estimate the size of the animal, and whether it was one or several who had stood on the bank. He hesitated over informing his companions of the find; in the end he said nothing. After all, they would be moving on very soon.

People often thought of the Late Cretaceous as the glory days and the end of the Age of the Dinosaurs, as Julian knew; but it was equally the dawn of the massive radiation of mammals, from a few very similar types into the diversified groups that would spread over the earth. It was these small mammals scampering through the forest, Julian's own ancestors, that interested him more than any other type of animal. Often he sat off to the side in the twilight hours, silently watching for the creatures.

The most common type was about the size of a kitten and gray all over. It would rummage through the dry leaves and then sit up and gnaw on a beetle, with an audible little crunching of the carapace between its teeth. But it was not until he found a dead animal, torn apart and surrounded by Hilda's pad prints, that he was able to study its teeth and guess at the genus.

"Alphadon," Julian said, poking at the remains with a stick. "A marsupial."

Yariko squatted beside him, peering at the little bundle of fur and blood, holding her nose. "How do you know it's a marsupial?"

"The pouch," he said, prodding the belly.

"I know that," she said. "But how could you know from a fossil. How did anyone know that 'Alphadon' had a pouch."

Julian pried open the tiny mouth of his poor dead *Alphadon* and pointed out the unique marsupial dental formula: three premolars and four molars.

Alphadon was not the only mammal that Hilda obligingly killed. Each time she plodded into camp with a bit of gray fur dangling from her mouth, they would crowd around eagerly and Dr. Shanker would coax it away from her, much to her chagrin. Eventually she stopped bringing her kills to camp, and Julian would find them scattered around the forest, chewed almost beyond recognition.

They also found several larger animals killed, it seemed, by *Troodon*; these kills, with their distinctive disembowelment, were sometimes only a few hours old. Fortunately, none were found near their camp.

In addition to *Alphadon*, Julian saw a rat-sized, arboreal creature that might have been a species of *Eodelphis*. It had a ringed hairless tail and a hairless face; it looked rather like a small opossum, possibly being the direct ancestor of the modern animal. The opossum, he knew, was a wonderfully successful holdover from the age of dinosaurs, very little changed. Rodents he did not see; they would not evolve for another twenty million years.

Once he saw a *Pteranodon*, an immense, rust-colored creature, gliding with stately grace, a small airplane; vilified in the movies, but in actuality a toothless, harmless eater of fish. It gave out a shrill, lonely cry, caught an updraft, and disappeared to the north with a flap of its wings. He never saw another; they were already on the brink of extinction.

But the most wonderful animal of all was a tiny, unimpressive mammal.

It was sitting on a branch, nibbling at a beetle of some sort. It could almost have been a squirrel, minus the fluffy tail. Julian was alone, walking back to camp, not even looking for wildlife at the time; but as soon as he spotted the animal he froze and stared up at it, and it stared back.

It seemed to know that Julian could not climb. The animal continued eating, turning the dead insect in its forepaws, peering down, curious but unafraid. Julian could see the very beginnings of a separation between the tiny thumb and fingers; and on the foot, which was clamped to the branch, the great toe was separating from the other digits. The creature had large, dark eyes, and its skull was not so low or narrow as that of the other mammals.

Purgatorius.

The meeting was so ludicrous and awesome at the same time that Julian felt dizzy. This creature might truly have been his direct ancestor, his grandparent removed by three million generations. He wondered what was passing through its tiny, slightly expanded brain; he could not quite read the expression in its eyes. Finally he asked it out loud, "Well, are you proud of me? I'm your own child."

At the sound of his voice the animal dropped the beetle, screamed at him, and then clambered away and disappeared into the forest canopy.

Dr. Shanker was also too restless to sit still. He busied himself transferring large rocks from scattered piles to the campsite. He built a more substantial wall on one side of their shelter and they felt a bit safer sleeping next to it. Once he came back from an exploration and claimed, jokingly, that he'd found a cairn.

"Looked just like one," he said. "Gave me quite a turn, when I first spotted it. Just the size to cover a human. Of course, it was covered with creepers, which is probably what held the rocks together. I didn't see any carvings though." He winked at Julian. "This invisible rescue party is impressing me less and less all the time. First they leave a message in unintelligible symbols, and then one of them is killed and buried. Besides, it's not like the ground here is too hard to dig a real grave."

Julian had long since gotten over his embarrassment at trying to see letters on a rock; he chuckled along with Yariko, although they kept the ongoing joke from Frank.

When moving rocks was no longer useful, Dr. Shanker took to exploring the island with Hilda, the two of them crashing through the bushes like a herd of edmontosaurs. Indeed, Julian grumbled, they made rather more noise than the wary herbivores. Chasing all the interesting mammals away, he complained; to which Dr. Shanker countered that mammals were clearly dangerous beasts who would attack at will: witness Frank's still swollen shoulder bite.

Julian found it vaguely annoying that Dr. Shanker spent so much of the day alone with Hilda, away from the camp and essential chores; although, to be fair, Shanker was quickly gathering long, straight branches for building a raft. At the same time, and quite irrationally, Julian felt a twinge of annoyance whenever he came back to camp and found Yariko and Frank so busy and amicable together.

It felt as if they'd been on the island for a lifetime, even though it was less than seven days; Julian wanted to be moving, fighting his way west, not waiting for his companions to play at making tools.

The second evening he came back to find Yariko fitting a short stick against a crude-looking bow. "Watch out," she said as he approached. "I might shoot you. I haven't figured out how to make it go straight yet."

With that she pulled back on the braided vine string and let it go with a snap. The stick went remarkably far, Julian thought; almost thirty feet. It also went sideways, and tumbled in the air as it flew.

"Well, it might land on an animal's head," he said, helpfully.

"Yeah, and anger it," Yariko said, with a sigh. "It's not just the lousy arrows. It's the string too. Those vines are no good." She turned back toward the shelter, and Frank.

"Yariko," Julian said, and she stopped. "Why do you sit in camp all day?"

She gave him a puzzled look. "Why do you wander around all day? Frank can't move far and somebody has to stay with him. We've been making things, useful things. He's working on a better axe and spears with stone heads." When there was no answer she added, "We need these things, Julian. We need to get going on that raft, too. There's nearly enough wood gathered."

"Then let's start it," Julian said, still feeling annoyed. "Let's get down to the water and start putting those sticks together. Surely you must have calculated it all out by now." He stopped, surprised at the acrimony in his voice. He hadn't meant to be angry.

Yariko looked at him for a long moment before speaking. "It's true I'm trained as a physicist," she said at last. "And that was useful when we first got here. But now. . . ." Her expression as she looked at Julian was almost an appeal. "Now, I'm not a physicist any more. I've had to redefine who I am, my essential self. You wouldn't know. You're as much a paleontologist here as back at the university. You can just be yourself. Sometimes," she went on, sounding sad now, "I'm not sure who I am now. And this thousand-mile journey just seems insane. What does it matter? We can't possibly make it."

Julian had to suppress a new feeling of panic. Yariko, uncertain? Yariko not willing to try? And if she gave up, before they'd even really started, what would happen to him? It was her strength he needed to keep his own hope alive.

"You're Yariko, and you're my friend, and Dr. Shanker's, and—Frank's, too," he said, feeling generous. "And if you give up, I'll give up and we'll just live on this island forever. Maybe it wouldn't be such a bad life." Seeing an ironical smile on her face he went on in a lighter tone, "And if I'm still a paleontologist, then you're still a physicist. Physics still existed back in the Cretaceous, you know, just like paleontology."

Yariko snorted. "Paleontologists study the past, my friend. This is the present."

Their conversation was interrupted by Dr. Shanker's voice, loud as usual. To his surprise Julian also heard Frank's voice equally loud. He and Yariko hastily turned back to the camp, wondering what could have happened.

Frank sat in his usual spot; he was clutching his VHF radio in one hand and his gun in the other. Dr. Shanker stood beside him holding the crutch, as if he had just taken it from Frank and helped him to the ground.

"Stop playing with that thing," Shanker was saying as they approached. "There-is-no-one-to-call. Unless you're trying to hail a T. rex. If you'd brought two of the things there'd be some use in it; but it's just a piece of plastic now. And stop grabbing the gun at every noise. You know Hilda's step by now."

Yariko opened her mouth but Frank spoke first. "It was worth another try. Going west is all well and good, but it'd be a hell of a lot easier if someone came out east to find us."

Before Dr. Shanker could respond Yariko took his arm. "It doesn't do any harm to try," she said. "Leave him alone."

"Somebody should take the Goddamn thing away from him. Trying to call a rescue party. . . ." Dr. Shanker turned away in disgust. Yariko followed him, trying to calm him down.

Julian looked at Frank. He had closed his eyes and leaned back to rest against the tree; but he was not quite close enough, and only one shoulder touched the rough bark. He looked tilted, unsteady, crumpled. He had dropped the gun but was still holding the radio.

Julian sat down beside Frank, feeling awkward. For a moment he drew in the sand with his finger, lazy spirals and shapes, not knowing what to say. He did not like conflict; he also couldn't understand Frank's strange mix of hard wilderness practicality and obstinate refusal to believe his radio wouldn't work.

"Frank," he began at last, and looked up.

Frank was already looking at him. "What will you do, in the future?" he asked.

Julian was taken aback by the question; in answer, he said the first thing that came into his mind. "I'll write up all my observations and publish them."

"Yes." Frank said. "In Nature. I know that one—that's a big science magazine. Only great discoveries are printed there. You'll have solved all the mysteries. People will look at fossils in a completely different way, because you've seen the animals, and where they live. Fossils from later periods will make much more sense, too."

He paused and shifted his position a little, using his arms to scootch himself back to the tree. Julian was surprised; he had never mentioned these thoughts, and had always assumed Frank wouldn't understand.

"And then there's Dr. Shanker, with his Nobel Prize," Frank continued. "This thing he's done. . . ," he gestured around them, "this'll get it for him, he says."

Julian waited to see what would come of Frank's strange train of thought. A security guard among three academics, scientists at that, might feel out of place in another situation, but Frank had been their leader in day-to-day survival, and he must know that. Julian couldn't imagine getting very far without him.

"You can walk around," Frank continued. "Explore. Think about physics, and fossils. Things you know about. I just sit here," he patted his leg and then the dirt beside him, "and wait for the predators. And think about the things I know. Fortunately," he added with a snort, "you listen to me most of the time."

"Yes," Julian said. "We probably wouldn't have survived without you. All the things you think of. . . ."

Frank had more to say. "You—with your dentition and your footprints, your albadonts and sucho-whatevers; Shanker with his

Nobel Prize and physics equations—and you think I'm crazy because I can't stop from trying the VHF now and then? If there's the remotest chance you're wrong about the time period, I want to be in contact with someone."

"You don't have to explain," Julian began, but Frank went on.

"How am I going to walk a thousand miles? I'm stuck here, unless someone else comes along and takes us back. Oh, I'll cross this sheet of water with you, if we can make a raft, but what about after that? And look at yourselves. Wandering around daydreaming about the future. What future? The fantasy one, where you're back in Creekbend and famous for your discoveries? Or the real one—here?"

They began constructing a boat that evening.

The work went remarkably fast, taking only another day and a half. All of them felt the urgency of their limited window of two months; and the grueling work and sense of purpose brought them together again in supportive partnership.

With a common project to focus on Yariko seemed unconcerned with who she was, and became the driving force behind the boat; Dr. Shanker stopped pestering the others and spent more time in camp, discussing the future. He and Frank had many an amicable conversation about boats and their boating experiences. Even Hilda seemed to perk up; she liked to run off with the branches meant for the raft, lolloping in circles while everyone chased after her.

As for Frank, he took apart his radio and made a clever little saw, just right for shaping small wooden tools, out of a roughened piece of wire stretched tight across part of the plastic case.

The raft was assembled from the straightest and most buoyant wood that Dr. Shanker had collected; Julian had earlier done test runs on every species of tree in the forest and settled on cypress. The vines that grew in the low areas proved strong enough for lashings. The design, Frank's brainstorm, was like a low log-cabin arrangement made of thick branches, with a thinner but tighter floor on top. This way, he pointed out, they wouldn't be sitting low in the water and vulnerable to crocodiles; also, the raft couldn't be capsized.

Frank also insisted they make the raft long and thin, for better steerage; and to cap it off, he designed a crude rudder, somewhat like a steering oar. He and Yariko became the master oar makers.

Neither Julian nor Dr. Shanker had the patience to stand and shave wood for hours at a time. Julian found that his arm hurt after only a few minutes and he was easily frustrated by the makeshift tools, although Frank's homemade axe wasn't bad. Yariko persevered: she made some crude paddles and a bundle of sharp arrows.

And, purely by luck, she discovered an excellent material for bow strings: the tough fibers from the long, thin tuber they'd been eating. The root had a stringy core that was almost impossible to cut before cooking, and these fibers when twisted together could take the tension of the bow. The range was short, and the bow strings broke after three or four shots, but it was a start.

When the boat was at last complete they loaded it down with everything they could think to bring, and dragged it through the mud to the water to test it. Frank remained in camp, alone for the first time, but busily engaged in fashioning some last-minute tools.

"What's that?" Yariko asked, pointing at the turtle shell in the boat. It was filled with small pieces of white stuff.

"An Alphadon skeleton," Julian said casually. "To take back with me. There aren't any absolutely complete ones. . . ."

"You put those bones inside my water pot?" Yariko glared at him. "You're going to carry some animal's bones a thousand miles?"

"Wait till he tries to bring a T. rex skeleton," Dr. Shanker said. "Whitney, you can have your old bones as long as you don't make me carry them. Now let's get this boat, if you can call it that, in the water."

They shoved the raft until it floated free of the mud, then climbed carefully onto it and paddled around while Hilda stood on the shore and barked. The crude floorboards were painful to sit on and the boat proved difficult to keep in a straight line, even with coordinated paddling. Dr. Shanker had been a kayaker in his youth, and tried to apply the same techniques, without much success. Julian knew nothing about canoeing or rafting. Finally Yariko clambered over them and took up the steering position, and by working together they were able to control their direction.

Dr. Shanker joked about how she and Captain Frank could sit in the stern together and plot the course, while the grunts paddled. Julian didn't mind being a grunt, but he wanted to be the one sitting with Yariko. He thought they should paddle around the island for a bit and do some exploring.

But Yariko insisted they could not embark on any expeditions without Frank, the true designer of the boat. "Never mind dangerous," she said. "It isn't proper. Frank ought to be on the maiden voyage."

"You just want the gun," Dr. Shanker said, meaning to be funny. He got a sharp look in response, and Julian knew a sharp reply was coming next. Suddenly irritated with them both, he peevishly agreed to turn back. After dragging the boat ashore they walked silently back toward camp, Hilda plodding at their heels.

Partway along the jungle path they heard a gunshot, and then a second one. The sound was so unfamiliar and weird in that primeval forest that they froze and stared at each other, confused. It took an instant for the obvious to sink in: Frank was in trouble.

They began to run. As they came near enough to see the brown wooden angles of the lean-to through the trees, Hilda stopped suddenly. She growled, her ears laid back and the hair bristling on her neck; but she seemed frightened. She sat down in the trail, looked up at Dr. Shanker, and gave a whining yelp, exactly as if she were trying to speak.

They hurried into the clearing.

Julian saw blood, everywhere, pooled in the uneven surface of the mud, spattered far up the trunks of the sycamore trees. He could smell it too, and it had a sickening, pungent odor.

Then he saw Frank, sitting in his usual position with his back against a tree. His stomach hung open like a glistening, crimson mouth and a wad of loose intestines lay in his lap. Scattered around him in the bloody soil were the two-toed footprints of *Troodon*, mingled and confused.

A long moment of sick horror went by, and then Yariko screamed, "He's still alive!"

Frank was looking at them.

Julian's horror only increased. He couldn't move, but Yariko suddenly sprinted toward Frank. She stopped about ten feet away from him, and stood with her hands over her mouth. "What do we do? What do we do?" she said in a high voice.

Frank looked up at her. "The gun," he said, and his right hand that lay near the gun twitched.

Julian's legs became unfrozen and he approached Frank, one slow step at a time, his mind still dazed. He stopped well behind

Yariko. Dr. Shanker followed him.

"What do we do?" Yariko said again, turning to look at Julian with her eyes enormous in a chalk-white face.

Dr. Shanker looked down at Frank. "Get it all back inside, and wrap him up," he said. "Before the thing comes back for him." But he made no move to touch anything.

Frank's head turned from side to side. "The gun," he said again. "Quick."

Julian was confused; did Frank want them to find his attacker and kill it? But why hadn't those two shots killed it? Frank wasn't one to miss.

"But how can he walk? How can he move?" Yariko said, still in that unnaturally high voice.

Julian knew, intellectually, that Frank would be dead in half an hour; but some part of his mind clung to the irrational thought that he could be bandaged, splinted, and helped into the raft to begin their journey. Of course he would recover. Of course he was going with them.

Frank's head moved again. "Can't," he said, his voice weaker now. "Dying. Want to die. . . ." his right hand clutched weakly at the gun again, ". . . before it comes back. Don't want to be eaten . . . alive. Too slow." He tried to push the gun toward Yariko; she stooped and took his left hand, and his eyes looked into hers. "You have to," he whispered. "Now. Then go, get out. There's too many. . . ."

Yariko dropped his hand and backed away. "Oh no," she said. "Oh no, you can't ask that, I can't." She bumped into Julian. "He wants us to shoot him!"

Frank's hand still plucked feebly at the gun.

"He's right," Dr. Shanker said in a hard voice. "He's dying. He doesn't want to be alive when those things come back. We can't stop them."

Julian stared at the gun, and then at Frank's face. Nobody moved.

There was a noise in the bushes.

Frank's eyes moved from one to the other. Then his hand grasped the gun with sudden strength and lifted it to his head. "Run," he said, and fired.

They ran. A sudden motion attracted Julian's gaze and saved

him from a close look back after the deafening shot. Four animals stepped out of the bushes. Julian had only time to see that they were bipedal and shorter than him, and to note the huge, sickle-shaped middle claw on the hind foot; then he was running harder than he had ever run, Yariko and Shanker were running, pelting for the water and the raft. Hilda ran ahead of them.

They tumbled onto the boat and shoved it off with long sticks.

The *Troodon* hadn't followed. They had a meal already. Julian, Yariko, and Shanker grabbed paddles and dug them into the water without a word. All their thought was to go, get away, far away so they wouldn't hear anything, would stop imagining. . . .

They paddled for a long time in silence.

The boat rode low in the water, weighted with three people, a dog, and everything they had put on it earlier in the day: the axe and four stone-headed spears, the turtle-shell pot, crude fern mats and wooden poles for making a shelter, and Yariko's bows and arrows. Hilda lay in the bow on a pile of mats, Dr. Shanker and Julian sat 'midships and paddled, and Yariko sat in the stern, alternately paddling and steering. The afternoon sun burned in their eyes. Fish swam past, shadows in the murky depths. Some were enormous, maybe reptiles, longer than the boat; but none threatened them.

They made gloomy company, silent except for the dipping of the paddles.

The mainland came slowly closer. At last it was there, and they entered the wide mouth of the river. The smell hit them like a wall: mud and decay, too much life, too much death. Green shores and dense jungle, silent in the afternoon heat, slid past on either side. Finally a bend in the river put their little island out of sight behind, and they were all relieved to be rid of it at last.

Julian paddled with smooth even strokes: dip, turn, dip, turn.

After a while Dr. Shanker said, in a conciliatory tone, "Good job at the helm. Frank couldn't have done better."

Yariko was silent a moment, and then said, "You didn't do anything. None of us did anything. We just stood there . . . you didn't even care."

Dr. Shanker's face was strangely white, his eyes still overlarge. But his voice was the same as always. "Of course I care. Do you think I want anyone to die, especially that way? And our chances aren't any better without him."

"Our chances? What about him? What about his chances? It's ended. It's already happened."

"Yariko," Julian said, in a tight voice. "Just steer. Don't talk about it. Don't."

He didn't want anyone to speak, he didn't want to hear anything; but she looked at him and said, "You're no better. You couldn't even go over to him. You couldn't even look at him."

Julian stared at her, unable to retort and equally unable to admit the truth of her words.

Yariko shoved the long steering oar to the left and the boat turned back toward the middle of the river, away from the too-near left bank.

"I don't. . . ." Julian began, and then stopped. He suddenly felt too weary to bother.

"Hush," Dr. Shanker hissed. "Keep your voices down."

"Why?"

"Because," he said, "you've already attracted a predator. Look— "

Part II

Hell Creek

One of the most productive fossil beds of the Maastrichtian is the Hell Creek Formation in Montana. But even here, the fossil record is incomplete. The fauna we know best are the shelled animals, because of their excellent preservation. The areas we know best are seashores, swamps, and rivers, because the sediment was more likely to cover up and preserve biological remains.
—*Julian Whitney*, Lectures on Cretaceous Ecology

– EIGHT –

1 September
2:54 PM Local Time

Chief Sharon Earles was pacing behind her desk again. "We have no leads," she said. "Not one lead on those missing people."

"And they're not the kind of people to go missing together," Hann put in, unhelpfully.

"What is it with the Cremora?" Agent Kayn was spinning himself in Earles' chair, his legs stretched out to keep his feet off the floor, watching in bemusement as Hann took half a cup of the chunky powder before adding coffee.

"He's on a diet. No cream or sugar," Earles said, with no hint of sarcasm.

Hann nodded and sipped the gooey liquid. "No doughnuts either," he said sadly. "How about that student's story?"

Earles shrugged. "Beetles. Top secret experiments. All well and good, if that's what he thinks; but I'm interested in the conditions at the time of the explosion, not in how exciting their findings were."

"But he thinks he can figure that out," Hann persisted. "If you'd let him in there to do it."

"I can't," Earles said. "He's not licensed to handle that equipment alone. He'll have to wait for the arrival of those two other physicists I contacted."

The phone on Earles' desk rang. It did so every few minutes, but so far little real information had come in. "Sharon Earles speaking." She put her hand over the mouthpiece and whispered, "Sergeant Moore, from Roscoe." This sounded more hopeful.

Agent Kayn nodded and flipped open his notebook. Hann looked up, a cigarette dangling forgotten in his hand.

"We'll be there in half an hour. Please touch nothing. Thank you." Earles hung up and looked at her companions for a moment. She wasn't quite sure how to present this development; it wasn't good, nor was it conclusive. And if true, it would negate her own growing suspicions.

"They've been located?" Hann asked hopefully.

"Not exactly." Earles watched her officer carefully as she spoke. "They found a body at a gas station. Male, dressed as a security guard, so far unidentified." She hesitated, but decided to leave it at that.

"Oh no," Hann said. "No, not in Roscoe. Can't be."

"Let's go." Earles ushered them out and into the car.

Sergeant Moore met them at the gas station. Several police cars were parked any which way among the dirty pumps and a small crowd of onlookers stood to one side, kept back by a young woman in uniform.

"It's not a pretty sight," Moore told them after the introductions. "He was killed within the hour, we think . . . in a rather chilling way. Can't think who could have done such a thing, or why. Never seen anything quite like it." He led them around the cinderblock building to a weedy lot full of old tires and broken glass. "Strange the way he just appeared, too. Nobody heard anything or saw anyone acting suspicious. Employee just came out for a smoke and . . . there it was, just like that. Gave the attendant the fright of his life."

Earles saw what he meant when they reached the spot.

Her first impression was of blood, everywhere, pooled around old bottles, spattered on tufts of weeds. It had a sharp, disturbing smell that made her want to back away. Then she saw the body. It was sitting up, leaning back against an enormous tire. One eye was wide as if in shock; the other wasn't there. A chunk of the face and head was missing from one side. The waxen features on the intact side were spattered with blood and looked inhuman, unrecognizable..

"Had his brains blown out," Moore said, in the tone of voice one might use to comment on the weather. "Lovely day today" could easily be the next thing out of his mouth, Earles thought.

The forensics man knelt beside the body and began to take rapid notes.

Hann suddenly appeared from around the gas station; he had lagged behind, unnoticed. Now he stopped a good ten feet away.

Moore pointed out the empty holster. "Think it's your missing guy?"

Hann spoke before Earles could. "That's not my brother," he said, and vomited into the weeds.

Earles sighed. She was glad this wasn't their missing security guard; glad for Hann. But that made it just another false lead, and time was moving on.

• • • • •

Dr. Shanker's voice was deceptively casual. "You've already attracted a predator. Look."

Julian looked at the shore. At first he saw only vegetation: branches and vines, tangled and mingled up with the undergrowth, shot with ferns and enormous tree roots that straggled into the water. Then he saw a face peering out of the jungle: a long naked snout, brownish-green and blending superbly with the background. The forward-facing, stereoscopic eyes of a predator glared at him.

"All creatures swim," he said in a low voice. "Almost all. But I can't see it taking the risk. It would have no chance in the water against a crocodile or a mosasaur."

"What is it?" Dr. Shanker paused in his paddling.

"I can't quite see enough of the body. A small theropod—a carnivorous dinosaur. Too big for Troodon, I think."

"There's another one." Yariko pointed. A second face had thrust out of the leaves, beside the first.

"Pack animals," Julian said. "Like wolves."

"Shall we pull away?" Dr. Shanker said, his paddle poised.

One of the animals pushed the twigs aside with a clawed hand and stepped to the water's edge, one foot settling in the shallow mud. It cocked its head and stared at the boat. It stood about the height of a human, and may have weighed a hundred and fifty pounds. The S-curved neck gave it the alarming air of a snake about to strike. Julian saw the enormous claw on the inner toe, curving upward, jutting above the mud.

"Dromaeosaurus," he said.

"A larger version of Troodon." Dr. Shanker reached for a spear.

Julian nodded, still clutching his paddle. "Like lions and cheetahs. They hunt different prey and keep clear of each other's territory. Although we haven't seen the real lions yet: T. rex."

"Good thing Hilda's looking the other way," Dr. Shanker said. "Let's get out of here before she spots them."

One of the dromaeosaurs opened its mouth and made a noise, a rapid clicking in its throat, like a low growl. At the sound Hilda started up and turned; when she saw the creatures she unceremoniously scrambled over Julian, knocking the paddle from his hand, and all but leaped off the side of the boat in her eagerness to get closer.

"Let's go," Yariko said. "We shouldn't tempt them by sitting here."

Julian turned his back and picked up his paddle. But as soon as they began to pull away he heard a frightful scream behind him, like a human in agony. There was a splash, and he jerked around to see what had happened. The creatures had leaped into the water, not just one, but three, swimming toward the boat.

Julian dug his paddle in harder, nodding to Dr. Shanker. "Paddle. We should be faster than they are." He couldn't help looking back over his shoulder.

They swam with their heads up and their long bodies and tails submerged just beneath the surface, stretched out behind them. He watched their three-clawed hands scooping at the water and their big hind legs treading; they made astonishing speed. The tail, he noted, simply trailed behind, being essentially a stiff balancing device, not very flexible. The dromaeosauridae in particular had bony rods, ossified tendons, running along the back end of the tail to stiffen it. Good thing, he thought: they'd move a lot faster if they could scull.

All these intriguing thoughts went through Julian's mind in the first few seconds, while the boat seemed to be pulling away from the animals. But once the dromaeosaurs reached deeper water where they could tread more easily they began to put on alarming speed. With real paddles and a well-made boat, the picture might have looked different; but even paddling as frantically as they could there was no chance of outpacing them. Julian kept turning to look back as he knelt and each time the animals were closer.

"Don't turn around," Yariko finally snapped. "I'll tell you what they're doing. You just paddle." She was steering into the current now, leaning hard on the long oar to compensate for Dr. Shanker's more powerful strokes against Julian's weaker ones.

Hilda, who had been keeping up a steady growl, now burst into a barking frenzy. She looked like she was about to launch herself into the water and attack them, in which case she would certainly have been killed. Finally Yariko had to let go of the steering oar and throw her arms around Hilda's neck to hold her back, and the boat began to yaw, swinging right and then left with the uneven paddling.

"We can't do it," Dr. Shanker said, tossing his paddle into the middle of the raft. "Grab a weapon. Let the Goddamn dog go and grab a weapon! She can take care of herself!"

The boat swung to and the creatures closed in on it broadside. Hilda stood on the edge snarling and barking, then dashed from one end to the other, precariously jolting them around.

Julian took up one of Yariko's stone-headed spears and prepared to stab at anything that came near. But even as his heart pounded, some part of him was still fascinated by the creatures' behavior. They paused just out of reach, and one of the three circled around to the other side of the boat. They were clearly used to attacking large animals in the water and followed a highly effective, coordinated strategy. Maybe they took the boat to be a peculiar variant of a hadrosaur or some other dull-witted herbivorous dinosaur.

They all attacked at once, screaming so frightfully that Hilda stopped barking and tumbled backward into the boat. Dr. Shanker grabbed the makeshift stone axe to fend off the creature on his side while Julian watched in terrified fascination; it was all he could do to turn his back on the fight and face his own attackers. He imagined Dr. Shanker overcome, and the creature silently clambering onto the boat behind him.

Then the other two came forward, their long jaws open showing jagged teeth, and there was no time for imagining the worst.

Yariko jabbed an arrow at one of them: the slender shaft suddenly sprouted from its shoulder, flapping about as the creature swam. It seemed to have no effect at all.

Julian thrust his spear with all his strength into the face of the nearest animal. He could tell by feel that he had struck flesh; but the animal wriggled aside. Three claws like grappling hooks caught the side of the boat. The animal was trying to climb onto the raft the same way that a human would, grasping the rim with its hand. It raised its head over the edge and screamed, jaws gaping, blasting the rotten smell of its breath.

It was streaming with water, and Julian could see a gash on the side of its neck where the spear had hit. Blood ran into the water. But the animal didn't seem to feel anything.

Suddenly it lunged upward to snap at Julian's face. He started back instinctively but the ragged jaws and putrid stench kept advancing, coming closer to his face. He fell over on his back and saw the head draw back and then dart forward with terrifying speed on the snake-like neck. With a yell, Julian rolled and found himself sitting upright; he was still clutching the spear, and with both hands he brought it straight down on the head, pushing with all his might.

The point struck the flat top of the skull in the center and drove the animal's head downward. Julian could feel the hardness of bone and he wanted to pierce its skull and brain, pin it against the bottom of the boat, but the animal wriggled free with a violent twist that tilted the raft and threw Julian over on his side. The spear slipped away and he caught himself with his head dangling over the side and his face almost in the water.

He struggled to regain his balance, expecting any second to feel teeth around the back of his neck. But when he sat up and looked, the animal was thrashing madly in the water, without coordination. The concussion had been fatal.

Julian slowly became aware of the others again. Hilda was cringing in the center of the raft. The animal that had attacked Dr. Shanker now lay still in the water, its head gently turning and twisting in the wavelets, connected to the neck by a few shreds of muscle and skin. Only Yariko was still fighting; she was holding the splintered shaft of an arrow, stabbing at the animal's open mouth. The broken end of the arrow was bloody, and both of her hands were red as if she had dipped them in a container of blood.

Julian's spear was floating in the water just beyond reach. He had a wild thought of lunging for it but instead he threw himself on Yariko and pulled her back, away from the dromaeosaur. They tumbled over as the raft pitched, landing on their backs, Julian still grasping Yariko by the shoulders. She looked back at the dromaeosaur and struggled to sit up.

But Dr. Shanker was already there with the axe. The animal seemed to realize that it was outnumbered; it backed away, hissing, teeth and gums red with its own blood. It swam for shore, the spear still flapping against its shoulder.

Yariko pulled free of Julian's grip. They watched the defeated animal drag itself out of the water and vanish into the foliage.

Julian felt no exhilaration or even relief as the animal disappeared into the trees. He felt an immobilizing hopelessness. They could never survive a thousand miles of this. They would not have survived this attack had they not been in the river; and most of their journey would be on foot.

After a long moment of silence Yariko spoke quietly. "I've been bitten." She held out her bloody hand: the creature had snapped at her and raked the skin from her palm. The wound was still oozing blood, and one gash was quite long and went down to the bone.

There was little to be done except to bind it securely with a handkerchief. At first she didn't seem to feel any pain, and as Julian awkwardly tied up the bandage she even joked about their brave and fierce dog who had hidden from the fight. But when she reached for the paddle to steer again, her face changed.

"You should lie down," Julian told her. But she only gave him a quizzical look, and he realized that there was nothing for her to do, one way or the other, but keep going on as before. In this world, one did not lie down after an injury.

"Whitney," Dr. Shanker said, interrupting suddenly. "What do you think?" Julian looked up. Shanker was holding up the neck of one of the dead dromaeosaurs, and the nearly severed head flopped about, leaving great wet blotches of blood in the hair on his forearm.

"Should I drag it aboard?" he asked. "There's some good meat on the leg."

"For Godsake," Julian snapped. "Get rid of it."

Dr. Shanker looked surprised and baffled. "It's good meat," he said.

Julian glared at him, and finally he shrugged and dropped it back into the water.

They slowly paddled away from the floating corpses, trying to keep to the middle of the stream. Yariko insisted on steering even though she looked quite pale and sick. She held the paddle with one hand and her wrist, since the other hand was too tender.

"Whitney," Dr. Shanker said, after quietly paddling for a time, "you keep telling us that dinosaurs are diurnal, mostly, and the little mammals are nocturnal. Why don't we take the hint?"

"Become nocturnal ourselves?"

"Why not? I know there're nasty things running around like dromaeosaurs, but can't we climb a tree and sleep until dusk? We'd make better time on the river if we weren't being attacked. And we need to make time: there's a thousand miles to go in less than seven weeks."

Julian scanned the shore doubtfully. He wasn't too keen on climbing out of the boat only to face another pack of dromaeosaurs. The woods seemed populated with the tall cypress, but the downward sweep of their branches made them a bad choice for sleeping in.

He was about to express disagreement when he saw some other trees up ahead: the sycamore-like trees familiar to them from their island, and clumps of shrubs that looked like small maples. The wide, multipronged leaves were unmistakable, although the plant was otherwise nondescript, nothing like its far-distant offspring. From this unlikely swamp plant, *Acerites multiformis* as Julian knew it from the fossil record, came the familiar sugar maple.

"All right," he said at last. "We should boil some water to clean Yariko's hand, anyhow. We can stay near the river and get in the boat if anything comes along."

They pulled cautiously to the shore. All was silent and still. "I miss the sound of birds," Yariko said as they stepped out into the stinking mud. "Even insects, crickets. There's nothing singing. It's unsettling."

"I'm not crazy about insects," Dr. Shanker grunted as he gathered sticks for a fire.

There was nothing to eat, and nothing they could do about it. Nobody relished the idea of charging into the undergrowth with a sharpened stick, looking for a small mammal to kill. They would have to sleep hungry, but at least they could drink their fill of boiled water if they could start a fire. Dr. Shanker got to work immediately.

Julian helped Yariko to sit on the end of a log half buried in the mud. She was obviously making an effort to be cheerful but it was equally clear that her hand hurt a great deal.

"That turtle shell really comes in handy," she commented after Julian filled it with water and put it over the fire, trying not to smother the flames completely. "I'll have to get one for my kitchen. I wonder if they come in different sizes."

But Julian couldn't laugh. He suddenly realized they had more to fear than carnivore attacks. Infection from a small injury would kill them as surely as would being torn open by *Troodon*.

The afternoon was well begun when they doused the fire, drank the remaining still-warm water, and munched distastefully on a few shriveled roots they'd brought from the island.

Dr. Shanker made a dry nest between two protruding roots of the nearest "sycamore," an ancient, gnarled tree of tremendous girth, and lined it with dead leaves. At his signal Hilda curled up into the space.

"Don't make noise," he told her, exactly as if she were a person who could understand him. "Lie quiet and let the predators pass you by." Then he scrambled up into the lower branches of the tree, with the help of Yariko's Julian's back. "Pass up some sticks from the boat," he called down. "There's a good spot here to make a small platform."

Julian took a moment to react; he was studying some curious markings on the trunk of the tree.

"What are you looking at?" Yariko asked.

With his finger, Julian traced some lines etched into the bark, a little above his head. "Reminds me of kids carving their initials in trees," he said. "You come upon them sometimes in the woods, in ancient oaks and maples that used to stand at the edge of a field. Carved with a knife, all distorted after decades of growth."

"What, you think dinosaurs carve their initials in trees?" Yariko said impatiently, and thrust a stick into his hand.

Julian handed the stick up to Dr. Shanker. "It just looked like something at first," he said, feeling foolish. "But wouldn't it be interesting if dinosaurs sharpened their claws on trees? No Cretaceous mammal could have reached that high on the trunk."

As Julian handed up the final piece, Dr. Shanker suddenly asked, "Whitney, are there bees here? In the Late Cretaceous?"

"It's not known if they go back so far," Julian replied, pausing with the stick in his hand. "Flowering plants are only just now spreading over the world, and there's some debate over whether bees are secondary to flowers. They don't start showing up in sediment and amber until the Tertiary."

"I didn't mean for a lecture," Shanker said. "I was wondering what just stung me."

"You've been stung?" Julian and Yariko stared up into the branches.

Shanker shrugged. "Unlucky day." He reached down for the last stick. "Wasps, maybe?"

"Plenty."

"Yes," he said, "that's what it was. Look." He pointed to the underside of a thick branch, where the dried mud tubes of a wasp's nest clung to the bark.

"Dromaeosaurs I'll face," Yariko said, standing beside the boat. "But I draw the line at wasps." She looked weary and unhappy.

Dr. Shanker meant to solve the problem in his own blunt way: by smashing the nest with a stick.

"Don't do that!" Julian cried, just in time. "They'll be all over us. Let's find another tree. So much for dinosaur initials," he said to Yariko, but she wasn't in the mood for humor.

They found another good tree forty feet away and a little back from the river. The sticks were again placed, and Dr. Shanker climbed down. "Your platform is ready," he said. "I'm sleeping with Hilda. I want to keep guard over her—make sure she doesn't run off by herself. There isn't room up there for three of us anyway."

Julian lay down cautiously, not quite trusting the crude planks. There wasn't much room; he tried not to be up against Yariko but there was no way around it. She lay with her back to him, her head resting on one arm and her tangled hair straggling over her face. The air was slightly cooler at their height above the mud, and the flies and the stench less bad; but Julian had trouble closing his eyes.

The events of the morning were not easy to forget. The death had been gruesome beyond belief, with Frank's strength of will being the final horror. He lay a long time trying to push it all away. He forced his mind to look ahead to their river journey.

An idea came to him suddenly and he lifted his head to look at Yariko.

"Are you asleep?" he whispered, almost soundlessly.

"No," she said, not opening her eyes. "Not yet. Be quiet."

"I thought of a name for the river."

"Mud River," she said. "Death River."

"There's a fossil bed from nearly this time and location: the Hell Creek Formation."

"Hell Creek?" she said. "That's a good name. I like it."

Then a few minutes later she said, "I'm sorry, Julian. It was such a terrible day. Frank—and then those animals. There's no hope, is there? We're all going to die a horrible death, and it'll be soon, too."

She was verbalizing Julian's own emotions immediately after the dromaeosaur attack. He sighed, reaching a hand out to touch her arm. She turned quickly and he withdrew his hand.

"I know," he said. "That's exactly what I was feeling, in the boat. I know we beat them off, but . . . but I don't feel that way now. Hopeless, I mean. Yariko, I think we can make it. In trees during the day, on the river at night; and then, when we leave the river the terrain will be completely different, and much safer. At least, I think so."

Yariko was silent for a long time. Julian couldn't tell if she believed him or not. He only partly believed himself, but he was trying. If they had no hope they might as well give up now.

"It's too late for Frank," Yariko said.

"What happens when we—when he died?" Julian asked, having a sudden thought. "Will he stay here or will he revert, and be found?"

Yariko turned her head to look right at him. "How can he revert? He's not in the right place. Now he'll never be." After a moment she turned back onto her side. "He knew he wouldn't live long without being able to walk. He knew from the beginning he'd be killed before having time to heal, and he tried to teach us as much as he could so we'd survive."

"I know," Julian said. "I understand."

"No, you don't," she said, and after a moment added, "You never tried to get to know him."

Julian closed his eyes wearily, feeling even worse, if that was possible.

Dinosaurs were not aquatic. However, a large variety of reptiles lived in fresh and salt water during the Cretaceous. These included plesiosaurs, ichthyosaurs, the monstrous mosasaurs, turtles (including one species that was as big as a small car), crocodilians, and champosaurs. In freshwater swamps and rivers, champosaurs in particular seem to have been common. These crocodile-like carnivores grew up to eight feet in length and had long, narrow snouts. The placement of the nostrils at the tip of the snout probably allowed the animal to float just below the surface, almost entirely unseen.
—*Julian Whitney*, Lectures on Cretaceous Ecology

– NINE –

1 September
3:30 PM Local Time

Three people stood in the doorway of the main room at the physics lab. Two, a rumpled-looking man and woman, were hesitant to enter, hanging back; the third, Sharon Earles, impatiently gestured them forward.

Marla Ridzgy had just flown across three states on an hour's notice, without a second thought. Forty-six, with short graying hair, wire-rimmed glasses, and smart city clothes, she struck Earles as a take-charge person. Claude Bowman was likely to find himself in the position of assistant under Ridzgy. He had a more hesitant manner; he was older, balding, with a pale wide face shaped like a pear, eyes that blinked too much, and a widening middle.

Both scientists worked on the cutting edge of particle physics, and both had setups similar to this lab's. Earles wanted to know what Shanker and Miyakara had been doing that morning, and what had caused the equipment to explode.

Nothing had been touched, except that the body was replaced by a chalk outline and the blood on the floor was turning black. Earles was annoyed to see the blood; her physicists didn't need to see it, after all, and it should have been cleaned up by now.

110

"Please, don't skip the vault," she said, as her guests shuffled around wearing the special slippers and gloves she'd provided.

"Yes. It's just. . . ." Bowman gestured toward the blood-stained outline.

"No help for it," Earles said. "The man was cut clean in half. Only the lower half was here. The rest was gone."

Bowman looked startled and not a little horrified. "Oh well, that's for the coroners, or forensics, or whoever they are. We're interested in the equipment."

"Oh I'll look, then." Ridzgy approached the vault, being overly careful not to step on the dried blood. Holding her nose against the by now nonexistent smell, she stooped down and looked into the vault. "There's less damage than I expected, from what you said," she commented. "In fact the instruments are mostly intact. Burned circuits and broken dials—more of a wiring issue now than anything." She stood and stepped back around the blood stain.

"We'll need to see the notebooks, of course, and everything on the computers," Bowman told Earles. "Can we pack things up and take them? This may take some days, if not more."

Earles had no intention of so casually releasing the lab's secrets, if there were any. "You may work here," she said. "An officer will be stationed in the hallway, for your convenience. We have rooms arranged for you tonight, when you're ready; tomorrow, we'll talk about what you need to do. Please disturb nothing but the notebooks, and of course the computers."

"What do you think?" Bowman said, when Earles had left.

Ridzgy shrugged. "No telling yet. If the vault wasn't sealed during an experimental run, there could have been vibrations. But an explosion powerful enough to rip a person in half should have destroyed everything in there; yet it's nearly intact."

Bowman stared at the half-person chalk outline. "But he wasn't in the vault . . . or maybe half of him was thrown out by the force?" He went closer. The small round door was hooked open against the wall; he unhooked it and swung it shut, then open again. "Strange," he said. "They told us the door was sealed shut and had to be pried open. Was it—"

"Sucked in tight by a vacuum?" Ridzgy finished for him.

Bowman sat down and looked at her. "Not an explosion: an implosion, inside the vault. A vacuum. Something set up a vacuum.

How do you suck air out of a sealed ten-foot-cubed room and turn it into a vacuum?"

Ridzgy was looking at an open notebook that lay beside the main computer. "There's something funny about these programming notes," she said, bending closer.

• • • • •

Julian woke suddenly from a sound sleep. His hip ached from resting on a knobby branch, and he squirmed around trying to ease it while wondering what had woken him. The sun was low in the west, and the air was hot, close, and damp. There were no sounds at all.

He rolled onto his back, vaguely worried about the total silence, and then he saw: he was alone.

As he sat up the silence was broken by Yariko's yell.

He scrambled down from the platform, showering the ground with bits of bark and twig in his haste.

Dr. Shanker had started up. "What is it? Who's that?" He jumped up and grabbed a spear that was leaning against the tree beside him. "Hilda?"

"It's me," Julian said. "Don't stab me. Yariko's in trouble. This way." He dashed into the trees.

He saw Hilda first: she was sitting near a clump of thorny bushes with her ears pricked forward and her tail slapping the ground, exactly as if she were waiting for a treat. He was reassured until he spotted Yariko, partly screened by the bushes.

She had a look of shock on her face that made his heart race. Julian wanted to rush forward and pull her away from the danger, whatever it was; but his legs didn't want to move. He could only stare back at her.

"Yorko!" Dr. Shanker's loud, practical voice made them both start. "Whitney! Don't just stand there in the bushes. If something's in there, get away from it."

Yariko shook her head and slowly came around the bush. Something hung from her hand; something smelly, and limp, like a dead animal.

"What did you see? What are you holding?" Julian stepped closer.

Yariko's face was white. She extended her arm and said, in a very controlled voice, "It's Frank's holster."

"What?"

Yariko held the limp black something up higher. "Hilda found it. I took it away from her when I saw what it was."

"What are you talking about? You're dreaming. Drop that thing and get away from there." Dr. Shanker was clearly annoyed; but his voice was not as steady as usual.

"Yariko. . . ." Julian didn't want to touch the filthy thing that he now saw must be a piece of animal skin; dinosaur or maybe reptile, he thought, from the absence of fur. Probably weeks old, black with rot. "It's just a bit of animal skin. Something died and Hilda found it."

This distraught Yariko imagining things was very disturbing. But Julian could almost understand. There'd been no time, the day before, to take in Frank's horrible death. It was still hard to believe; he still expected to hear Frank's voice every moment. Yariko must have climbed down from the loft, half asleep, and coming upon Hilda chewing these ancient remains, immediately imagined something familiar that belonged to Frank. Her face was flushed now, her injured hand swollen; she might be feverish. No wonder she was so upset.

"This is not animal skin," Yariko said, still with that strained look and tight voice. "It's part of a plastic holster made to look like leather. It's Frank's holster."

Julian was relieved when Hilda leaped up and snatched the thing out of Yariko's hand. As the dog ran off joyfully with her prize he took Yariko's arm and steered her back toward their tree.

Dr. Shanker helped sit her down with her back against the tree, and they made her drink some water. Julian cleaned her injured hand and rebandaged it. It was red and hot, and Julian had a sudden fear that the infection was spreading, making her delirious. But Yariko didn't seem to notice. After a few minutes she blinked up at him and then looked around her.

"Are you awake now?" Dr. Shanker asked.

Yariko scowled at him. "I've been awake for a while," she said in her usual voice. "But . . . maybe I was seeing things. It couldn't have been, could it? Of course not. How could Frank have gotten here? For a minute I thought, I imagined that he'd been following us." She took a deep breath and looked at Julian. "I'm sorry I frightened you. Maybe I wasn't quite awake. I was dreaming about him trying to follow us, trying to catch up. . . ." She closed her eyes and leaned back again.

Julian shuddered. He'd been having the same dream, not surprisingly. It didn't bear remembering. "You curl up here and sleep a bit more," he said. "We'll keep watch."

"I want to know what you were doing climbing down by yourself," Dr. Shanker said. "Wandering off into the trees without telling anyone."

"I climbed down to get some water," Yariko said. "I wanted to fill the turtle shell and try to get clean. I meant to be fast but then I saw Hilda with that thing.

Dr. Shanker snorted. "Next time you get it into your head to take a bath, let one of us—preferably Whitney—know about it. None of us should go anywhere alone."

The afternoon was declining; a drizzle began as the light fell. They lit a fire with some trouble and once again dined on hot river water. Julian closed his eyes and tried to push off the hunger pangs. His stomach felt pinched and caved in. How long had it been, now? A day and a half? They'd munched on some raw roots, stringy and unpleasant, but hadn't managed to find a solid meal since leaving the island.

"We should catch something before we get too weak to hunt," he said, remembering Frank's advice.

"What about fish?" Yariko looked more herself after her nap and a good scrubbing of her face in clean water. To Julian's relief she seemed to have forgotten her strange imagining. He only hoped Shanker wouldn't add this to his "rescue party" joke.

"Need bait to catch a fish," Dr. Shanker said. "Let's get back in the boat. It's going to be dark soon anyway."

They paddled in silence for while. The sun eventually went down although they didn't see it leave; the world simply became black, and the drizzle became a light rain. Julian didn't know what to think of the river. It had seemed the perfect means of travel; but they couldn't catch the nocturnal mammals for food if they were out in the middle of the river every night. Now, if they could catch something larger and somehow preserve the meat. . . .

From behind him Yariko whispered, "I hear something." The jungle was silent, except for the hushed and vast background sound of a slight wind in the leaves.

"A splash," she said. "It was very close. You didn't hear it?"

"Some carnivore, I'm sure," Dr. Shanker grunted.

Julian put out his hand and rested it on the top of Hilda's head. "She's looking to the right," he said. "Hilda. I think she heard it too."

Then they all heard the splash, and something wet smacked into Julian's lap and was gone in an instant. There was a scuffle in the boat, and then Dr. Shanker hissed, "Let it go! Hilda! Let it go!"

"What is it?" Yariko said.

"I've got it by the leg," he said. "Should we cook it for a snack? I'm hungry enough to try one, anyway."

"What is it?" Yariko repeated.

"A frog, a big one," he said with a laugh. "We could get a mouthful each out of it."

Even two days ago Julian would have gagged at the thought of frog; now he reached for it. "Let's cook it," he said. "It'll be better than nothing."

"Wait—it's going to escape. . . ." Dr. Shanker seemed to be scrabbling around again. "Let go! Hilda! Too late. She's got it."

They could hear her crunching happily for the next several minutes, while they thought about their own hunger.

"At least we can get clean," Yariko said as the rain became even heavier. "And in the dark, too." Julian didn't know what that meant at first; then from behind him he heard the unmistakable sound of a zipper.

"What are you doing?" he asked, rigidly facing straight ahead although it was impossible to see anything at all.

"I'm taking a shower—a real one." Now Yariko was pulling off her jeans. The boat jerked about as she tried to steer while getting undressed. "You should do the same, you know. Spread your clothes out on the raft to get rinsed, and let the rain fall on you."

Foolishly, Julian felt himself blushing. Yariko was right, and he felt as eager as she to find some semblance of cleanliness. But undressing in the dark when she was right behind him, only three feet away . . . with nothing on . . . he pictured her kneeling by the steering oar and blushed again.

"I'll let the rain soak through," Dr. Shanker said. "It works for Hilda, so it should work for me."

"It's really lovely," Yariko said, her voice sounding very close to Julian. "Warm and cool at the same time, and I can feel the mud sliding off." The boat swerved suddenly. "Sorry—just putting my arms up for a moment. Ah." Then she laughed. "I don't hear any rustling of clothes from forward," she said. "Come on, Jules. Don't be childish."

Julian thought his feelings were anything but childish; but he gave in. He couldn't have her thinking he was that shy. If it had been just him and Dr. Shanker, or any other woman, for that matter, he wouldn't have hesitated an instant. He tried to make as little noise as possible.

Yariko was right, he decided once his crusty jeans and reeking shirt were off. He heard her laugh again, and the subtle sound of skin rubbing on skin. He cupped his hands for water and then scrubbed at his bristly cheeks, his neck, his shoulders. It did feel glorious. The water trickled around his nose and dripped off his chin; it went under his arms and down his chest and back almost like a caress on his dry skin. He couldn't believe such a simple thing could feel this good. It was sinfully, almost lustfully sensual.

Something bumped his shoulder and he jumped back.

"Sorry," came Yariko's voice, sounding startled. "I didn't know I was so close."

"OK you two water babies, pool's closed," Dr. Shanker broke in. "We're drifting downstream, losing ground. Time to paddle. In any case, I'm getting a little embarrassed. I can't see what you're doing in this pitch blackness."

Julian felt a shocking surge of desire as images came into his mind. For a moment he couldn't stop them; he wondered if Yariko was having similar thoughts. Then he came back to reality. If she was picturing anything, it wasn't his face that was close to hers, or his body. She'd be thinking of that mysterious fiancé, who Julian pictured something like Frank; and at that thought, his desire slipped away.

Yariko could be heard getting into her clothes; but Julian stayed as he was for a while longer.

After some time the rain stopped and the clouds opened up. The moon went down. It rested a moment on the treetops like the huge face of a celestial dinosaur, and then disappeared. The stars that showed as the clouds retreated were more brilliant without any competition from the moon; they were so bright that the outline

116

of the river could be seen again, where the faint speckled reflection of the sky ended and the black mass of the jungle began. The air felt warmer, and less close.

"Do you know," Yariko said suddenly, "they aren't the same stars. I wish I had studied my astronomy. The Crab Nebula, for instance; a huge glowing cloud of gas blown off by a supernova. But it hasn't happened yet. That gets to me, somehow." After a few moments of silence she added, "But then, is North America even in the same place on the globe? Could that explain why the stars are different, and why it's so warm?"

"No . . . we're in the same latitude now as in our own time," Julian said. "About forty-five degrees north, though not as far west; the Atlantic Ocean is young yet, and only just beginning to open. As for the climate, it's significantly warmer now all over the globe. Atmospheric CO_2 is incredibly high—we're in a real hothouse, with almost no polar ice caps. That's why sea level is so much higher."

When the sun rose, mist began to curl from the water. A few birds fluttered in the leaves along the river, some of them dropping plummet-like from a branch and disappearing with a splash, after the fashion of a kingfisher. There were no song birds and the silence of the sunrise was uncanny for those used to the morning din of calls and trills.

All around in the quiet water and the slow current, fish began to rise for insects. They'd hear a plop of water and then see the rings spreading out. They tried to shoot toward the center of the rings, taking turns with the best working bow, but they were hardly quick enough. Each time the arrow would plunge in and then bob back up again and float on the surface, clean. They stopped after every few shots and paddled quietly about to collect the arrows. Julian was ready to groan with hunger.

Suddenly there was a disturbance in the river about thirty feet ahead. "Look!" Yariko cried, pointing. "Something's going after the fish."

The something was hard to identify, until it finally caught a fish in its jaws. Then Julian saw the narrow snout with protruding nostrils on the very end, and the exceptionally long thin teeth as they impaled the struggling fish. "Champosaur," he said.

"Looks like a crocodile to me," Dr. Shanker said, backpaddling. "How big is it? Will it attack us?"

"They're small," Julian said. "Four to eight feet at most. I don't see why it should attack us when there're so many fish. Look," he went on. "It's leaving a trail of pieces. Maybe we can scoop some up as bait."

The water was now calm; they slowly paddled closer and Yariko reached over with the turtle shell and scooped up the floating debris.

"Don't get your hand near the water," Julian cautioned. "The champosaur might still be there."

They managed to pick up three shreds of what looked like internal organs. Yariko took apart one of the bows and Dr. Shanker turned it into a very crude fishing pole and line.

The first piece of bait was pulled right out of the loop they'd tied around it, and they never even saw what got it. Then Dr. Shanker took off his watch, tore off the thin metal tongue and clasp, and with some effort bent it into a tiny hook.

Almost instantly something bit. Shanker yanked it in so hard that he went over backward on the boat, and the foot-long fish never had a chance to let go; in fact, it was all they could do to keep Hilda from grabbing it as it flopped around on the raft.

"Try again," Julian urged, greedy in his hunger. "Let's get two."

They did. The second one took longer, but it was almost twice as big. It got a bit mangled being dragged onto the raft; Yariko was worried a champosaur would detect the blood and come after them, so they hastily paddled for shore.

In the bright dawn with the mists still dissipating, they cleaned the fish and made a stew that seemed a feast. The last of their shriveled roots rehydrated as the water came out of the fish and a lovely smell went up. Nothing had ever tasted so good. Julian didn't care that his fingers and face were sticky with fish.

"Good thing we didn't decide to bathe in the river," Yariko said thoughtfully, licking her fingers. "That champy-thing had pretty long teeth." She kept a number of bones for fishhooks. Dr. Shanker saved selected fish parts as bait, carefully wrapped in large leaves and set in the crook of a tree.

They were able to construct a wider and more comfortable platform than last time, and Yariko even pulled some leafy fronds and sent them up, to cover the worst of the bumps. However, Dr.

Shanker still insisted on sleeping below with Hilda. They curled up side by side in the boat, which had been dragged almost completely ashore. A hoary, homespun string was tied loosely around Hilda's neck, and the other end to Dr. Shanker's wrist, to wake him if she moved.

As the sun climbed up above the trees, the jungle grew lighter, and the loft was flecked with little bits and chinks of light falling through the leaves. The air smelled clean although the morning breeze felt chilly. Julian and Yariko lay side by side, not quite touching. After a while Yariko turned away from him, settling on her side to sleep as she usually did. Her shoulder touched Julian's and he pressed in closer rather than moving away.

When she didn't react, he lifted his head and gazed at her. Her hair, still slightly damp, was in a tight shiny braid. Her T-shirt was cleaner but the jeans were still mud-spattered, torn and already bleaching in the subtropical sun; one hand was bandaged and puffy, and there were several scratches on her arms and face: she looked nothing like the Yariko of the physics lab. Yet, he still saw her the same way: decisive, competent, and calm. Dr. Shanker might see himself as the leader; but Julian felt that if anyone's characteristics were to bring them to their goal, it would be Yariko's.

"It's starting to sink in," he said, lying back again but still with his shoulder against hers, "that we're the only people in existence. Us three. That's the whole of the human population. Before. . . ." He hesitated, worried about upsetting Yariko, but decided to go on. "Before, I was thinking we had to get back home. I mean, I thought of the world as full of humans and our creations and society, and us just temporarily isolated from the world. But now. . . ." He couldn't tell if she was even listening.

"What?" Yariko asked, without turning over. "What do you feel now?"

"Well, now I understand that this is the world; that there are no people. They don't exist. There is no world of humanity to go 'back' to. There's only this world—our world." Julian stopped. He wasn't making much sense, but the feeling was strong, and sudden. It was an entire change of outlook, of philosophy even. But it was not a sad feeling. It was just different.

Yariko turned her head to him and laughed at a sudden, ridiculous thought. "We could populate a village, you and I."

Julian felt himself turning red. "Maybe a little one," he said. They seemed to be thinking in parallel, but she was ahead of him; too fast for him, even as a joke. "It would be a very inbred village," he said at last, reverting to science.

"Of course," Yariko said. "Isn't that common in small biological systems? You of all people; you shouldn't be squeamish. All of our children would sleep with each other perfectly wantonly, and produce more children."

"For Godsake."

She was thoroughly enjoying his discomfort. "But how can you be shocked? We're Cretaceous animals now, aren't we? Part of the wilderness."

"Maybe Dr. Shanker could contribute to the gene pool," Julian suggested.

Yariko made a face. "Who's being disgusting now?"

"I don't mean with you," Julian quickly amended. "Our daughters."

"Will he still be virile in fifteen years?"

"Knowing him, yes."

Yariko looked thoughtful. "Maybe we'll produce a whole race of people. Do you know, I'm not joking. Maybe we'll spawn an entire civilization."

Julian thought about it a moment, while brushing away a fly that landed on his arm to drink from the sweat. "No," he said, "it wouldn't be possible. There needs to be a critical mass to spawn a population."

"Well, two men and one woman seem like a critical mass to me." Abruptly, Yariko's smile was gone. "Unless two of us get killed, and the other is left all alone. . . . One of us could be the only human in existence. One of us could live a whole lifetime in solitude, if the other two died." She looked so stricken at the thought that Julian actually reached out an arm and drew her closer.

"We won't die," he said, fighting down thoughts of Frank. "We're all healthy. We're making good progress. I certainly don't intend to leave you all alone—or be left without you." He tried to speak lightly, but his arm tightened around her.

"Besides," Yariko said, and her voice took on its teasing tone again, "you paleontologists would have found traces of it in the fossil record. Stone buildings, or something."

Julian shook his head. "Not necessarily. All of our written history, six thousand years, is nothing, a blip, a blink of the eye, geologically speaking. All of those layers of rock could easily have swallowed it up, and we might never have found the slightest hint of it."

"Then it's still possible," Yariko said, with satisfaction, and without pulling away from his encircling arm. "A great civilization: immense dynasties and cataclysmic wars that would put to shame everything we were ever taught in grammar school. Don't you think?"

He was beginning to catch her enthusiasm. "Yes. And they would live in the higher, stony ground away from the Inland Seaway. They'd build steep hills with hollow insides, stone forts to keep out the carnivores. And they'd round up some of the smaller herbivores for cattle. They would use tyrannosaurids as war steeds. Bring them up from hatchlings, and train them to be loyal. That's an impressive thought—pounding over the open terrain on a beast like that."

"How exciting," she murmured. "Tell me more about it."

He tried to invent more, about great seagoing ships, imperial hunting expeditions to bring back prize dinosaurs, slave pits with dromaeosaurs instead of lions; but after a while he realized that she had fallen asleep. Her eyes were closed, and her expression was perfectly contented.

The Amazon Jungle is not well understood. Many of the species are either so rare or so remote that they are completely unknown to us. The small plants and insects, the foundation of that immense and complex system, are so numerous and diverse that scientists cannot hope to catalogue the half of them; certainly not before the jungle is destroyed by agriculture and everything in it becomes extinct. How, then, can we understand the ecosystem of the Late Cretaceous jungle, extinct for 65 million years? One thing we know: large animals would have been rare. You might have walked for days without seeing a single large, photogenic, scary carnivore.
—*Julian Whitney,* Lectures on Cretaceous Ecology

− TEN −

1 September
5:23 PM Local Time

Earles closed her door for a moment's quiet time. But she'd hardly reached her desk again when Hann entered without knocking, the two physicists on his heels. All three were bursting with news.

Earles was amused in spite of herself. "Who's first?"

The woman—Ridzgy?—spoke first, as Earles had known she would. "Preliminary assessment," she said, flipping open a tiny laptop and seating herself, uninvited, on Earles' desk. "Here's the sequence we're proposing. One, the initial explosion that you say was reported by the guard. Then, within sixty seconds of that, a large vacuum was set up inside the vault."

"A vacuum?"

"A sudden drop in air pressure, so that the door was pulled shut and sealed itself; an implosion, if you like."

Earles nodded. "The door was quite difficult to open when we first got there."

"Yes. It would be. Of course we weren't there to measure the vacuum, but the readings in the vault were recorded in the master computer. We're studying them now."

Earles sat down behind her desk, forgetting in her interest to tell

Ridzgy to get off. "OK. What caused the first explosion?"

"Unknown, so far," Bowman said. "Such things can happen if there's noise in the vault; for example, if the door wasn't sealed and a loud noise occurred—"

"Especially if a certain small but integral part is used in the alignment," Ridzgy interrupted.

"What part? Alignment of what?" Earles asked.

"Well, we can't be sure yet. The initial explosion shattered the dials and indicators, and as you know it's a bit of a mess. However, there's every indication that they were using a magneto-electric buffer to allow for micro-fine adjustments of the field."

"In English, please," Earles said, looking at Hann. "And without the caveats and hesitations: just give me your line of thought, substantiated or not."

Ridzgy nodded. "OK. This may be a long step, but here it is. There's a small platinum bar in the loop from the fine adjusters to the wiring. A piece of metal say as long as my palm. It's a kind of standardizer, or constant, that subtly modifies the electrical field. Kind of like the iron balls on each side of a ship's compass allowing fine adjustments to the magnetic field, only not so crude. It allows some very delicate control over settings; but it also makes the equipment supersensitive to noise and vibrations. I've never actually seen a setup that made use of it, until now."

Earles was thinking hard. "Is there more significance to this thing than just making an explosion more likely? Could it be related to the implosion, this drop in air pressure? And where did the air go?"

"Don't know yet," Bowman said. "It's not as if you can grab a parcel of air and just . . . vanish it, leaving a temporary vacuum. We're studying the results now to see if they recorded the volume of air that was, well, moved."

He seemed to be looking at Ridzgy for approval as he spoke. For that matter, he seemed to be looking at her most of the time. Earles was mildly annoyed; she wanted two independent opinions, not two people voicing the same thoughts because one needed the approval of the other. Or maybe there was something else at work?

She shook her head and went back to relevant thoughts. "There was a graduate student involved with their work. I'll send him to you tonight. He seems to know what they were up to. But he didn't mention a platinum bar, so maybe it's not important."

"The platinum bar may have been significant," Ridzgy said. "It could have amplified their settings in unexpected ways. In any case, a lot of power shot through that system: not only are circuits burnt, but that bar is partially melted. And platinum has a very high melting point."

"How high?"

"Almost eighteen-hundred degrees Celsius."

"Which is. . . ." Earles tried to do the calculation in her head, but Ridzgy was quicker.

"Over three thousand degrees Fahrenheit."

Earles sat very still. So that's it, she thought. That's the answer. Those people were vaporized, carbonized, whatever it was called, in a three-thousand-degree crematorium. She turned to look at Hann.

"Charlie," she said, and paused. Years of practice giving empty words of sympathy didn't make this moment any easier. "I'm so sorry."

Hann looked startled. "What?" He held yet another Styrofoam cup in his hand, probably loaded with the false sweetness of powdered coffee creamer.

Earles took a breath, and then closed her mouth. Not only did she have to express awkward sympathy, genuine though it was, but she had to give it to a man who was oblivious. "The vault was heated to three thousand degrees, Charlie," she said as gently as she could. "The people inside . . . I'm sure they felt nothing. It would have been instantaneous."

Hann's cup hit the floor with a soft splat. The sticky liquid sprayed his crisp trousers and shoes. He stared at Earles as if he thought she had killed his brother herself. Then he stood up and walked unsteadily to the door, crushing the Styrofoam cup underfoot and leaving the wet outline of his boot heel with every step on the dirty linoleum.

Earles let him go. "I'll get that coroner in again, to look for remains . . . ash, I suppose," she said, and she wondered irrelevantly if human ash could be distinguished from canine ash.

• • • • •

As the Cretaceous autumn advanced, the weather often turned to drizzle. Fortunately the seasonality of North America was subtle in that time, so it was still fairly warm. At first the rain was welcome;

drinking water didn't have to be boiled if they could catch it in the turtle shell as it fell, and staying just short of filthy without having to try river bathing was a plus.

The travelers fell into an easy routine, sleeping during the gradually cooling days and continuing up the river at night. The champosaurs were left behind, or became shy; no large terrestrial carnivore disturbed them. In fact, no large animals at all appeared for many days. They began to relax from the constant edge of desperate fear. Julian's interest in spotting new dinosaurs revived, and they made good progress on the river; they estimated close to thirty miles a night against the slow current. At that speed, and even with the windings of the river, they might make a thousand miles well before the time ran out. But they hurried all the same. Delays would be inevitable, and it was too much to expect that the river would cooperate the entire distance. They did not often talk about the subject; it was more than enough keeping up with the moment-by-moment demands of each day and night.

Food seemed more abundant, too, or maybe they were becoming more adept hunters. Yariko was the first to bring down a small mammal with a bow and arrow. Julian was impressed; he hadn't really taken her bows seriously before. After that, they shot one nearly every evening before setting out on the river again. Dr. Shanker and even Julian became reasonable shots as long as the target was perfectly still and no more than twenty feet away—not an impossible happenstance, as it turned out, the animals not knowing enough to be frightened of humans. Not only did they eat well, but Julian was able to study many kinds of mammals and began to note subtle differences between them.

Placental mammals from that period were all very much alike, and especially among insectivores the fossils were so similar that even defining genera was difficult, much less species. Julian was fascinated by the thought that from such conservative creatures, small brown eaters of whatever they could find, came the whole panoply of Tertiary mammals: the great cats, mastodons, ruminants, sloths, pangolins, cetaceans, bats, rodents, primates. It made him wonder: of all the inconspicuous animals known to twentieth-century zoologists, which ones would diversify into new species, while the others disappeared? What would life be like, sixty-five million years after the time of *Homo sapiens*?

Once they woke up in the middle of the day in a pouring rain. Julian sat up, his hair streaming and the drops drumming on the platform. The light was so gray that he could not even tell what time of day it was. The rain had knocked down leaves, which stuck to his skin uncomfortably, and the water felt cold.

Yariko was already awake. She was sitting on the edge of the platform, water dripping from her hair, and dripping off her feet that dangled over the edge. She gave him a small smile and said, "You look like a drowned rat."

"You look like a drowned primate," Julian countered. He felt awkward; once again they had fallen asleep with his arm over her, as had become habit, but it didn't seem to mean very much. Yariko needed his companionship as he needed hers, certainly, but he felt no greater emotion coming from her. And until he did, he was not going to push his own feelings on her—not when they were all so vulnerable. Yariko sitting up in the rain looking cold and unhappy, rather than choosing to curl up against him for warmth, made him doubly unhappy.

They agreed that it was impossible to sleep, and climbed down to see how Dr. Shanker was taking the rain. They found him squatting in the boat with Hilda; several of the palm-leaf mats were propped up on sticks and tented over his head. He wasn't much drier than they were. The boat was in a deep puddle, and the air inside his tent was heavy with the smell of wet dog.

Together they dragged the boat through the mud to the river. When they climbed aboard and took up positions, they were covered in mud as well as being wet and cold.

The opposite bank was almost unseen in the gray rain although the river had been steadily narrowing as it went west. Seen from out in the middle of the stream the banks took on a ghostly quality, appearing and disappearing in the grayness. The water seethed and hissed in the heavy rain. It was difficult to tell if they were making progress, because they couldn't see any landmarks to judge by. The makeshift floorboards were black from the wet, rubbing against the bottoms of their legs and causing rashes. Hilda curled up in the bow in a profound depression and refused to raise her head all day. Eventually Yariko propped up some of the frond mats over their heads, but the whole wobbly structure fell overboard and floated away.

For several hours they paddled miserably, not talking, hardly even looking up. Finally Julian raised his head to shake the sopping hair from his eyes and caught a glimpse of something moving on the shore.

"Stop," he hissed, suddenly attentive.

"Is this your Ornithomimus that you keep talking about?" Dr. Shanker said irritably. "Fine time he picked to show up."

"Quiet." Julian stopped paddling and the boat swung around. "I don't think it knows we're here." *Ornithomimus* was a peculiar and rare theropod about which very little was known. He longed to see one.

At first he thought it might be *Ornithomimus*, but as the boat drew closer he glimpsed its face for an instant. It was chewing on wet leaves. It must have been a hadrosaur, because of the wide duck-bill shape of the mouth, but there might have been a crest rising up from the back of the head; it was hard to tell. Then the rain closed in again and the bank disappeared.

"Did you see the crest?" Julian whispered in excitement.

"What crest?" Dr. Shanker said. "What animal? I can't see a God-damn thing."

"I saw the animal, but I didn't see a crest," Yariko said. "What did it look like?"

"A long, tubular crest, curving backward, on the head. You didn't see it?"

"No," she said, but she admitted the view wasn't very clear.

"Parasaurolophus," Julian said, picking up his paddle. "For good-ness sake, let's paddle to the shore and get a better view of it."

"What's the significance of it?" Yariko turned the steering oar toward the bank.

"Whether it's still extant. It was once very common, but the crest-ed hadrosaurs by and large died out by the late Maastrichtian."

The river was shallow near the bank and the bow of the boat stuck in the ooze on the bottom. The leaves stirred; the creature must have snuck away into the forest at their approach. Julian was disappointed, but he stared intently into the dripping foliage, hop-ing to catch a glimpse of it running away. He saw nothing except the wet jungle and the rain.

Yariko crawled over to him. "Are you sure it wasn't just a branch, sticking up behind its head?" she asked.

127

"No," Julian said sadly. "I'm not sure. I'll never know now, will I."

"Julian," she said, putting her hand on his shoulder, "there're too many mysteries for you to solve them all."

He leaned over the edge of the raft to inspect the ground where the animal had been standing; but if it had left any footprints, they were already turned to mud and puddles in the rain.

Dr. Shanker's voice startled him. "Whitney! We're not pulling to the bank every time you want to look for a dinosaur. We'll be eaten alive. Start paddling!"

Julian never again saw anything resembling *Parasaurolophus*. If it lived at that time, in the higher floodplains of the river, it must have been either very rare or exceedingly shy.

Dr. Shanker picked up a stone and threw it at the underside of the loft in the tree. He had settled on this method of waking up the others, although Julian could never tell if he was being discrete or was too lazy to climb the tree. If discrete, it was not necessary: he and Yariko had still done nothing more than sleep and talk together on their private loft.

Julian sat up, swung his legs over the edge, and peered down bleary-eyed at the ground. The sun was high and the forest was suffused with the greenish light of midday. "What's up? It's still broad day." He wasn't happy about being woken.

"Julian," Dr. Shanker said, peering up anxiously, and Julian knew instantly that something was wrong by the use of his first name. "Hilda's gone. She doesn't answer, she's not anywhere nearby. I think something's happened to her."

Julian ran his fingers through his damp hair. "What do you mean, gone? When did she go?"

"I don't know," he said. "I woke up and she was missing."

"She's exploring," Julian said, still grumpy from the early wake-up. "Leave her alone."

"She should have been back by now," he said. "I've been waiting for a long time. I'm telling you, something happened."

Julian sighed and began to pick his way down the tree. He'd expected something like this to happen eventually. For weeks now Hilda had stayed quietly beside them, only wandering short distances, and it was too much to expect of a curious dog in a fabulous new place.

As they'd progressed inland the ground had become more dry under foot. Now they were surrounded by climax forest with a firm ground, carpeted with dead leaves and shot here and there with ferns and mushrooms. It was ideal ground for an active dog to explore. She had probably followed a scent trail into the forest and was beyond earshot. Julian had little hope of finding her, unless she returned on her own—if she could.

They inspected the ground everywhere around the tree, and saw nothing to show which way she had gone.

"I used to think she could always take care of herself," Dr. Shanker said, "but with some of the animals here, I'm not so sure. She's very fast, though. I can't believe a two legged animal could possibly outrun her."

Yariko poked her head over the edge of the platform. "What's happening?" she said. Then glancing around, "Where's Hilda? Did she run off?" Julian nodded, and she climbed over the edge of the platform and down to the ground.

"I should have kept her on the rope," Dr. Shanker said. "Talk about delays. We'll never reach the mountains in time."

"Don't be silly," said Yariko. "Dogs don't get lost that easily. When did she go?"

He explained again.

It seemed a poor idea to wander through the forest searching for her, since she could have gone in any direction. Neither did it seem wise to split up and call for her, but that's what they did: Yariko remaining by the tree at Julian's insistence while he and Dr. Shanker went in opposite directions along the bank for a short way, calling for Hilda.

Julian realized how little he'd been using his voice lately when he became hoarse by the tenth yell of "Hiiiildaaaa!" Nervous of leaving Yariko alone, he soon turned and made his way back downstream to their tree. Yariko wasn't there; she was twenty feet away bogged down in a tangle of undergrowth, calling for Hilda.

Dr. Shanker soon joined them; he had seen and heard nothing.

The sun turned toward afternoon as they looked at each other.

Yariko lifted her head suddenly. "What was that?"

Julian strained to hear something; then the wind changed and he caught the sound of barking, very faintly. It was coming from deep in the forest.

The history of mammals is primarily about teeth. Reptiles, excepting a few snakes, have simple teeth designed for slashing and tearing. Mammals, however, have extraordinarily complex and specialized dentition. Many different kinds of teeth exist in the same mouth: incisors, canines, premolars, and molars. They are capable of precision cutting, stabbing, grinding, and even snipping by means of a shearing action, exactly like scissors. If you have ever seen a cat chew a tough piece of meat on the side of its mouth, then you have seen the marvel of carnassial teeth. The mammalian capacity to process food quickly probably underlies our high metabolic rate; that is, we may well have achieved our present ecological importance by means of superior dentition.
—*Julian Whitney*, Lectures on Cretaceous Ecology

– ELEVEN –

1 September
7:16 PM Local Time

The Creekbend police station was bustling. Early evening was always a busy time, and today even more so. The forensic man, Agent Kayn, had just left his final report; that OSHA man had been in again with his little twitching mustache, telling Earles things she'd found out hours ago; Mark Reng had been located and sent to the lab; and Hann had been hanging around off and on, red-eyed and fiercely fending off any expression of sympathy, trying to pick up bits of information.

At least she was able to give him some positive news when he came in next. "There was no ash or any other . . . remains in the vault," she explained as he paced her office, smoking a cigarette down to a nonexistent stub. "Nothing. Only those few drops of blood, a lot of fingerprints, and some dog hairs. They weren't burned to nothing, or vaporized." When Hann's expression didn't change she went on, "Frank may still be OK. There's no indication that he was killed in the lab."

"But you said that room got up to three thousand degrees," Hann argued, as if he was afraid to begin hoping again. "That would have . . . they would be dead. All of them."

"Yes, but they don't seem to be. The forensic man says there would be some remains, you know; if nothing else, bits of metal from their clothing, and. . . ," she meant to say 'teeth,' but changed her mind. At the moment Hann was a bereaved relative, not an investigator. "It's difficult and frustrating not having any clues when people have disappeared," she said. "And you've gone through a lot in the past few hours, thinking about your brother."

"Half-brother," Hann corrected her. "Different father."

"But still your brother. Charlie, why don't you take the night off, and tomorrow too? Take as long as you need. Just keep remembering we haven't seen any evidence that Frank isn't alive."

"What am I supposed to do, sit in front of the TV all night with a pack of cigarettes?" Hann's belligerent tone was an indication of his worry. He never spoke to his chief that way. "I want to stay here and work with you. I have to know what happened."

"All right. It was just a thought, in case you needed time to yourself," Earles said, wondering what she would do with him. "For now, you could take all these reports and go over them somewhere quiet, and see what you can come up with." It was busywork; Hann was unlikely to come up with any new ideas. But he needed something to do and she needed her own time to think.

After he left Earles shut her office door; but now Marla Ridzgy was on the phone again sooner than expected.

"There are some indications here, in the notebooks," she said. "So we may be able to follow the program adjustments they made. But—"

"Did that graduate student I sent over know anything?" Earles interrupted.

"Not really. Whatever they were up to, they kept it secret even from him. But we've found some puzzling things. Perhaps there's another physicist we can discuss this with? The details are technical."

"I'll come over," Earles said, and hung up. She grabbed her belt and opened the door, calling "Hann!" into the corridor. "We're going to the lab," she said when they met in the entranceway. Earles was a little annoyed at the physicist's tone. So the woman didn't think the police chief would understand "the details." Well maybe I won't, she thought, and Hann certainly won't. But we'll get the gist all right, and contribute some ideas of our own.

Bowman and Ridzgy were seated side by side in front of the master computer, intently watching lines of code scroll by. Earles waited politely for the sequence to end before speaking.

"Have you found the cause of the vacuum?" she asked.

"No, that's still unknown. Readings show that the air pressure in the vault dropped suddenly, after the sequence had been run. Don't touch that," Bowman added, taking a laboratory notebook away from Hann. "We've got them all in order."

Hann shrugged and put his hands in his pockets. "What does "apparent spatial translocation" mean?" he asked, leaning casually on the counter.

Ridzgy gave him a sharp look. "It means just what it says: an apparent change in location of an object. This notebook is filled with calculations and measurements for some very strange objects." She opened the notebook again and pointed out a particular page. "See here? Pebbles, beetles—why beetles, of all things? Twigs . . . weighed, then placed in different parts of the lab and observed."

"Sounds pretty boring to me," Hann commented.

"What did these beetles and twigs do while being observed?" Earles asked.

Ridzgy and Bowman looked at each other and then up at Earles. "They disappeared," Ridzgy said at last. "Just like your missing people."

• • • • •

"That's her," Dr. Shanker said after pausing to listen.

It was difficult to pinpoint the direction, especially since the slight breeze distorted the sound. But it seemed to come roughly from the south, straight back from the river. Then the breeze died, and it was silent except for the buzzing of flies.

Dr. Shanker looked back toward the forest with a grim expression. "You two stay here, in case she comes back."

"Don't be an idiot," Yariko said, and there was fear in her voice. "We can't separate. We'll never find each other again."

Julian agreed with her. "We have to stay together—always." He went to the raft, half hauled out on the bank, and began tossing their possessions onto the shore. "Spears, bows, and arrows," he said. "Enough for all of us. What about this?" He held up the turtle shell.

"Too heavy," Yariko said. "Anyway, we'll be back here before long. Right?"

Julian didn't answer. He glanced once more at the raft: it looked like a child's creation, rough and crooked, uncomfortable in the extreme, almost laughable; but it had been their home for nearly a month, and he felt strange leaving it for a walk in the forest.

"Too bad we don't have a ball of string," he said.

"String? That wouldn't hold Hilda." Dr. Shanker grabbed a spear and a bow.

"No, but we could loop it around trees as we went, so we wouldn't get lost," Julian explained. "It's what I used to do as a kid. However, this might do just as well." He hefted the makeshift stone-headed axe. "We'll mark the trees as we go."

"Come on then," Dr. Shanker said.

They set off into the forest, walking south. Every minute or so Dr. Shanker called out Hilda's name, but his voice seemed to fall on the dead needles underfoot, unable to penetrate the vast stillness of the forest.

Julian wasn't the only one feeling nervous about the forest. They'd spent the past few weeks traveling at night, seeing only the moonlit shores of the river, silvered leaves of trees, and a few mysterious black openings into the forest. Now they peered through the trees in uneasy curiosity at this new land. Julian was surprised at how much the terrain had changed. This was clearly a totally different ecosystem, far removed from the sea. Indeed, they must have been five hundred miles from the Inland Sea by now.

All around the huge trunks of sequoia rose like pillars. When Julian stood still for a moment and gazed up, the trunks seemed to converge at the top because of their enormous height. Shafts of sunlight fell through, touching the pillars and bringing out the red tint of their bark, or by luck piercing all the way down to the floor of the forest and lighting up a patch of ferns. At the base, the trees were so large around that one could have hollowed them out and made a house inside.

They were not modern redwoods; Julian knew this because they had begun to drop their leaves for the winter and the ground was cluttered with wide brown needles rustling underfoot. True redwoods are evergreen trees. These were evidently Dawn Redwoods, *Metasequoia*, wonderful trees known from the fossil record. They were famous because scientists had assumed they were extinct until an isolated grove of them was found in central China in the 1940s. Saplings were dug up and transplanted all over the world, and they soon became common ornamental trees. However, he had never before seen an entire forest of them. It seemed almost a sacrilege to mark them with the axe, as he did nearly every tree.

One intriguing speculation about why trees evolved such great height was that it protected their leaves from herbivorous dinosaurs. Certainly, Julian thought, the Dawn Redwoods would have been beyond the reach even of the great sauropods such as *Apatosaurus* and *Diplodocus*. However, those giant herbivores no longer existed in the north. Somewhere in the Southern Hemisphere the titanisaurids may still have lifted their unbelievable necks to reach the tops of trees; there had even been a Late Cretaceous migration to as far north as Texas, bringing sauropods to North America once again. But in the northern regions the sauropods had long ago disappeared, leaving behind a permanent mark on the floral morphology.

Julian was jolted out of his fear-tinged awe at the forest by the unmistakable sound of barking. It was faint but clear: the kind of bark a dog makes when it has treed something, a sharp, short, endlessly repeated bark.

"That way!" Dr. Shanker said, pointing a little to the left. He began to jog, calling "Hilda! Here girl!" every few steps. Yariko kept just behind him, and Julian brought up the rear with some vague chivalrous notion that a surprise attack from behind would kill him and not her.

Then the barking stopped. The forest was eerily silent.

Julian was still unsure of this venture away from the river, and how far it might take them. Even putting a dent in every tree they passed, which greatly slowed down their progress, didn't ensure against losing their direction. And surely Hilda had heard them by now—they'd been able to hear her, after all. She would follow if they headed back.

"What if we stopped and let her come to us?" he suggested. "I don't think we should go any farther from the river. I'm sure she'll find us when she's ready."

But Dr. Shanker strode on. "Just a little bit further," he said. "We've hardly gone two hundred yards; the river's only just back there." Suddenly he stopped and pointed. "We've walked up on something," he said, casually.

Yariko whispered, "Is it a rock, or an animal? Julian—take a look."

About a hundred feet away in the trees, up a rising slope in the forest floor, lay a gigantic gray object that certainly resembled the huge rocks left behind by glaciers. But the region had been free of glaciers for at least a hundred million years; so far they'd seen no such rocks. They watched it in silence for a minute, but it did not move.

Dr. Shanker thought that Hilda might have trotted up to sniff it. They all had the same unpleasant thought: that she had been killed, and the enormous creature was resting after its meal. Flitting from one tree to the next, they climbed the needle-strewn slope, until they were thirty feet away and too nervous to peer around the giant bole of the tree to see what it was.

"Stay here," Dr. Shanker mouthed; he was afraid even to whisper. He darted from behind the tree to another one nearby, gripping his spear so hard the muscles stood out on his arm and shoulder.

Suddenly he laughed and strode into the open. "Here's a fierce creature!" he said. "Come look!"

It had been dead a long time, and little was left except a mummified skin pulled tight across the bones. When they walked around to the other side they saw that the skeleton had begun to separate, and the vertebrae of the tail were scattered about on the ground. The feet were still enclosed in leathery skin; the leg bones had fallen under their own weight and lay partly sunk into the soil. It must have lain down on its side to die, and now the ribs arched up above their heads, gray and weathered, black in parts where they had begun to rot.

The size was impossible. To Julian, that was the best word: impossible. People of the twentieth century are used to regarding the elephant as the largest of land creatures, but an elephant could have bivouacked inside the arch of this immense rib cage. The skull was nine feet long, resting on the ground, its vast eye sockets gazing into the forest. It had a narrow, pointed face tipped with a horny beak.

Farther back in the jaw, exposed in death, was a battery of flat teeth like a grinding mill. A bony frill jutted back from the skull and yard-long horns curved out from above the eye sockets. It had a third horn, a smaller one, above the narrow nostrils. It was of course *Triceratops horridus*. It came over Julian finally how tiny a creature he was. The trees and the dried-up corpse seemed from another world, of vast dimensions perfectly proportioned for each other, and his own presence was ludicrous. He was a product of the diminutive world of the Quaternary, scampering through the haunts of these creatures of more perfect size. He expected to see a moth the size of a crow, or a cockroach as big as a dog, or some other proof that he'd been dropped into a world with a different scale of measurement altogether.

"What now?" Yariko's voice was low, almost a whisper, and Julian knew that she too had been struck with the feeling of insignificance.

It was time to get back to the river, and they all knew it. They hadn't heard Hilda's barking for some time, and without it there was no point in searching for her. If she was alive, she'd have to find her own way back to them. Still, they sat down beneath a great trunk and waited in silence for what seemed a long time, listening.

Dr. Shanker rose and turned back first, to Julian's surprise. He and Yariko followed, watching Shanker's slumped shoulders, and tried to cheer him up. Hilda must have come this way, Yariko said; she could never have passed a dead animal without sniffing it and this ultimate of dead animals must surely have attracted her like a magnet; and they all laughed at the thought of her innocent and funny curiosity. But inside they all knew that either she would come back to the river by herself, or they would never see her again.

Late afternoon crept in; Julian began to feel thirsty, and became careless in looking out for the notched trees.

Now and then they passed a clump of low bushes, bracken, and undergrowth living in the glow of indirect sunlight that filtered down between the giant trees. It was by the merest chance that Julian looked back over his shoulder for a second glance at one of these thickets, and saw a little patch of black fur.

The remains of a small mammal, he thought; somebody's meal. Maybe even Hilda's recent kill.

He snuck toward the bush, hoping nothing dangerous was about to come crashing out; but it did not move at all, and he decided it was dead. He pushed aside the twigs and peered down at Hilda, covered with blood and absolutely still.

The ceratopsia, the parrot-beaked dinosaurs, diversified in the Cretaceous. The coincidence of their rise with the rise of flowering plants suggests that they evolved specifically as the dominant eaters of angiosperms. However, by the end of the Maastrichtian, all except a very few species had died out. The massive Triceratops, *for example, was a late survivor, probably traveling in large migratory herds.*
—*Julian Whitney,* Lectures on Cretaceous Ecology

– TWELVE –

1 September
7:48 PM Local Time

In the physics lab, Earles stared at the two scientists. Physicists taking observations of beetles and twigs, beetles disappearing, physicists and dogs disappearing; she would have laughed if there wasn't a dead body to show it was quite serious.

"Come again?" she said.

"They disappeared," Ridzgy repeated. "The beetles did. Just like your missing people."

Hann snorted, but Earles jumped on the emphatic statement. "What do you mean, 'just like' our missing people?" she asked sharply.

Bowman made a dismissive gesture with his hand. "She was being dramatic. Of course we don't know where your people went."

Earles wasn't ready to let it go, however. "You said before there was a decrease in pressure inside the vault. What if some of the contents of the vault disappeared? Wouldn't that cause the vacuum?"

"It could have. If it was enough."

"Maybe as much as four-and-a-half people and a German shepherd?" Earles folded her arms and looked at Bowman.

"That's certainly one hypothesis," Ridzgy said.

137

Hann gave a bark of laughter. "What is this, Star Trek? People don't just 'vanish' into thin air."

"No," Earles said, before the physicists could answer. "But a few hours ago we thought they'd been burned to nothing; cremated. That would be a large amount of mass disappearing. What if they disappeared some other way?"

Hann stopped chuckling and looked impressed.

But Ridzgy shook her head. "Sorry, that doesn't work. Cremation, as you put it, isn't a negating of mass; it's a change in state. The water and carbon in the body simply become gas: carbon dioxide, hydrogen. The mass is still there. Furthermore, according to the notes and programs, these people were pulling small things into the vault from elsewhere . . . and then watching them disappear later, outside the vault."

"We're not here to solve the mystery for you," Bowman added, seeing the doubtful look on Earles' face. "We're physicists—we can try to re-create their experimental run, but there's no information here on people chopped in half or gone missing."

But Earles was not convinced. She was staring at the vault, seeing in her mind's eye the human body sliced through the middle; not burned, but sliced . . . cleanly, and the rest gone as if never present, blood and all.

• • • • •

Hilda lay partly on her chest, her head stretched out on the ground and one front paw curled under her body. Her lips were drawn back as if they had stiffened in a snarl, or perhaps a yelp. By now Julian had seen many dead animals, limp furry bodies with the warmth gone from them; but this death wasn't the same kind. There was a stillness about it that struck him like a blow in the stomach. This animal, their companion, so full of life and health and intelligence, this personality he'd lived with for so long on a tiny raft, this ever-curious and alert being who put the humans' adaptive abilities to shame; the physical proof that this spirit was suddenly gone, leaving only the final stillness of any empty body, a meaningless shell, made Julian lose all hope of his own survival. First Frank, now Hilda: the two who were best equipped to survive had been killed first.

He felt his companions approach and stand behind him.

Dr. Shanker dropped to his knees with an involuntary groan.

He reached out a hand toward Hilda's head and made a gesture of stroking it; but he didn't touch her.

She must have been badly injured, and crawled into the bush to die. As they stared at her it came over Julian that this was the natural end of every wild animal, including themselves. They were looking at their own future, probably not too distant, and the vision they had of reaching their vague goal was mere idealistic prattle.

Dr. Shanker pushed Julian aside. Then he gently disentangled Hilda's body and laid her in the open, as if, even in death, she might be more happy in a comfortable position. He squatted and palpated her body to see how she had been injured. The blood came from her mouth, and was coagulated in the fur of her muzzle and on her chest. It looked as if she'd been struck hard on the head or across the jaw. One side of her face was badly swollen and the flesh around the eye had puffed up and forced the lid closed.

Dr. Shanker felt gently near her collar bone, his stubby fingers now stained with blood. After a moment he looked up and said, "Her heart is beating."

Julian's new sense of doom vanished. He knelt down beside Shanker. "Are you sure? It's easy to feel your own pulse in your thumb."

"No," he said. "Try it."

Julian worked his fingers into the fur and immediately realized that she was warm and her body was not stiff. He could feel a faint pulse, and when he stared hard at her abdomen he saw a flutter, a hint of a breath, irregular and infrequent. She was just barely alive. "Yes," he whispered. "I feel it."

"If we let her lie still——" Dr. Shanker began.

"We should get back to the boat," Yariko said. "Before nightfall. It isn't safe here, for her or for us. And she'll need water."

"Not to mention us," Julian said. "I could do with some water myself." It was amazing how life suddenly mattered again, and thirst could drive him to action. "There's more likely to be food near the water too."

Dr. Shanker handed his spear and bow to Yariko and then gently gathered Hilda in his arms and lifted her. They walked back toward the river, single file again, slowly so as not to jostle her over the uneven ground. But after only a few minutes he stopped and said in a low voice, "There's something ahead." He lowered Hilda to the ground, and Yariko handed him his spear. "Do you see it?" he asked.

Julian looked hard, trying to estimate distances. Something gray showed far away between the tree trunks. It must have been huge, although his eyes were constantly tricked about depth and size in the gargantuan forest.

"Maybe it's a live one this time," he whispered. "Triceratops."

"We could make a wide circle," Yariko said, "but we'd lose our marked trees. And it'd probably hear us anyhow."

Julian gave her an apologetic look. "We've lost them already." He gripped his spear. "I'll go closer and take a look."

"I'd go with you," Dr. Shanker said, "but I want to stay with Hilda. If it's anything dangerous, she won't be able to defend herself, or even run away."

"All right, then," Yariko said, getting her bow ready. She pointed to a tree halfway there and said, "Not beyond that point. I won't let you. Frankly, this whole forest scares me. We don't belong here—and every animal will know that."

As they approached, they made out a gigantic gray object lying at the top of a gentle slope. They were within fifty feet of it and were crouching behind the tree, when Yariko suddenly said, in a disgusted voice, "Julian, it's the same one!"

It was.

They stood and looked at the massive skeleton of *Triceratops horridus.*

"It's not possible," Dr. Shanker said as he joined them, gently laying Hilda on the ground again. "How did we lose ourselves? We walked straight north toward the river, didn't we?"

Yariko shrugged. "Clearly not. It isn't easy to tell the direction of the sun, in a forest."

Julian looked wildly around at the woods, trying to see his crude axe marks on the trees, or some indication of the river's direction. Massive trunks rose everywhere, and deceptive lanes between the trees seemed to open up and then fade out farther along. Every direction looked the same.

"It can't be far from here to the river," he said, finally. "If we head northwest, we should hit it pretty soon."

They set out again, but more than an hour later still had not come to the river. The sky was hazy and it was impossible to tell where the sun lay. A diffuse greenish light filtered down through the treetops.

Yariko stopped first. "We need food," she said. "And water. Hilda must be very dehydrated. And we're not far from being in her state."

Julian knew they'd never reach the river before sundown, and it would be foolish to continue at night, in the pitchy blackness of the forest. They settled down at the base of one of the enormous *Metasequoia*. There was no possibility of climbing the tree, and no undergrowth to hide in; they would have to spend the night exposed on the forest floor.

They laid Hilda gently on a bed of needles. Her breathing had become more regular, and her foot twitched once in her sleep, a sign that the blow to her head had not paralyzed her.

Setting aside their spears, bows, and bundles of arrows, they settled back against the trunk of the tree, shivering in the slight chill of the evening.

"I wonder what it was that got her," Yariko said. She sat close to Julian, not quite touching, with her arms around her drawn up knees.

Julian had been pondering that. "My guess is she was chasing a group of herbivores, perhaps hadrosaurs," he said. "Like a dog chasing cows, or sheep: she was probably nipping at their heels and one of them kicked back. If she doesn't wake in the next twelve hours. . . ."

"She's strong," Dr. Shanker said. "She'll make it." But his voice didn't sound quite as assured as usual.

As the darkness closed in they fell silent. Julian began to doze, now and then waking up with a start to the same quiet forest, now invisible. Tiny sounds became more apparent, as did the scents: pungent needles of trees, resin, dry leaves, decay. Once he thought he heard a footstep, and he started awake and listened intently. He had just decided it was part of a dream when it repeated itself: a tiny thump. Both Yariko and Dr. Shanker remained asleep, their heads lolling back against the shaggy bark.

Again, the thump. It did not sound like a very large animal. Julian was reminded of a deer stepping slowly through the forest, barely audible, browsing on patches of ground cover. There was a long silence, several minutes, and then the sound again, much closer, directly behind the tree. He thought he heard the soft tearing of plants being pulled up from the forest floor.

He was bursting with curiosity as well as apprehension. It was clearly not *T. rex*, and didn't sound like any other dangerous predator; but a herbivore did not mean there was no danger: Hilda was proof of that. After all, many modern man-killing creatures were strict vegetarians: bulls, rhinoceroses, hippos. However, if his companions continued motionless and nonthreatening, they would probably be safer than suddenly jumping up and running off, or doing anything else that would startle the creatures.

Finally the animal came into view: not one, but a small herd, eleven of them, ambling one by one from behind the tree. They were bipeds, small ones only about the size of a human, or of a large deer, and Julian supposed they served very much the same function that deer might: nibbling on bark, bending down to tear up ferns and herbs, eating the leafy vines that clung to the tree trunks. They were comical animals, stocky, bull necked, and stout bodied, with heavy tails that jutted out stiffly behind and allowed them to balance on their two legs.

One peculiarity was so distinct that he knew the type of dinosaur instantly. The tops of their heads were expanded into large rounded domes, mostly bone, giving them the look of sage old men who had gone bald. Pachycephalosaurs.

They wandered into view peacefully, chewing bits of fern that dangled from their mouths. One of them pulled up the low herbs with its hands and then transferred them to its mouth.

It was Yariko who scared them away. She woke up with a start and said, "Good lord! What are they?"

They froze, staring at the humans like dinosaur gnomes caught in the middle of some piece of mischief. Then they vanished into the forest, thudding over the dry ground.

Yariko stretched, and gave a little laugh. "I'm glad we looked for Hilda," she said, her voice soft but startlingly close. "Now we're all together again we can find the river, and go on. You know, I think I'm getting used to this life. I can hardly remember living any other way. It seems so . . . normal."

Before Julian could answer, she had snuggled up against him with her head on his shoulder. Soon she was asleep, breathing softly.

Later in the night Julian heard the same thumping of feet. Another group of pachycephalosaurs must have passed close by, but he could not see them in the absolute blackness under the trees, and in any case he was more interested in what was sleeping in his arms.

When Julian woke next there was a dim light around him. Yariko's head was still on his shoulder, and his arm was around her; her knees had tilted over onto his lap. He hated to disturb her; but after a while a cramp in his leg demanded that he move. Yariko woke with a start, drawing quickly away from him. Then she smiled and leaned against him again.

"I forgot we weren't on a loft," she said.

Julian longed to kiss her on each of her puffy eyes, and on her very chapped lips; he wondered if she was wondering why he never initiated anything. He had always been so shy with her, but they needed each other now. Everything had changed between them. But even as he had the thought Dr. Shanker sat up beside them and stretched.

Seeing her master move Hilda lifted her head and wagged her tail. They were ecstatic. They crowded around, petting her and talking to her, and she seemed to like the attention. But when she tried to stand up she wobbled and then sat down suddenly and whined.

There was no water for Hilda, and no breakfast. They set off as quickly as they could toward the river, Dr. Shanker carrying her again. Her head lolled against his shoulder, and her wide brown eyes gazed back at Julian, sad and confused.

An hour later they still hadn't come to the river, or seen any of their notched trees. Julian silently cursed himself for losing them. It had seemed such an obvious marker at the time, but had turned out to be too subtle to keep track of easily.

Hilda began to struggle and complain about being carried, and when Dr. Shanker set her on the ground she puttered along at a maddeningly slow pace. The forest began to seem like an evil force keeping them from knowing their direction. Although nobody said anything, they were all thinking of the few weeks remaining before the close of the time window.

Again, it was Yariko who stopped first. "Now we really need food," she said in a faint voice.

Julian chose an arrow. "Let's get what we can," he said.

They moved more slowly, trying to be quiet.

The first thing they brought down was nothing more glorious

than a toad. Dr. Shanker got it with his spear. He fed it to Hilda and looked the other way as she devoured it raw. A little food in the stomach seemed to perk her up, and she was better able to trot along after that. But she seemed chastened, a little frightened by the forest, and showed no interest at all in wandering off.

"What's that?" Yariko said suddenly, stopping and putting her hand on Julian's arm. "Over there—something moving. Maybe several things."

She and Julian crept closer, trying to stay down, while Dr. Shanker waited beside Hilda.

It was a group of pachycephalosaurs; perhaps even the same herd they'd met during the night. One of the animals stood off on its own, separated from the others by a few yards, its head down, grazing on a patch of leafy trailers. Julian didn't dare whisper, but with signs pointed it out to Yariko. They split up, Yariko moving to the left, Julian to the right.

The pachys had chosen a patch of forest that was grown up with leafy, evergreen bushes, which fortunately gave cover for stalking them. Julian was able to get within about twenty feet of the lone animal without detection; there was no breeze in the still forest, and the animal was clearly not expecting an attack. He extracted an arrow from the bundle, silently, and fitted it to the bow. Then he peered out between the leaves.

The animal was turned away from him, and he waited for it to expose more flank, having no illusions about his marksmanship. The pachy took a step and bent down again to tear up a mouthful of leaves. But it must have heard something; suddenly it paused and stared over its shoulder directly at the spot where Yariko was hiding.

Instantly she released her arrow; Julian let his go almost at the same time. It sank into the animal's side just above the hip, while Yariko's went into the lower chest.

The creature screamed, a horrible sound, while chewed leaves tumbled from its open mouth. The whole group of pachys turned and ran, the injured one at the back and lagging.

Julian crashed out of the bushes to follow but they disappeared as if they had never been there at all. "I know we hurt it," he said, after a moment. "We should be able to track it."

Dr. Shanker joined them. "We'll have to trail it until it dies. Hilda can find it for us, I'm sure."

But Hilda wasn't much help when they brought her to the spot. When Dr. Shanker put her nose to a spatter of blood on the ground she licked at it and then glanced up with a wag of her tail, as if waiting for more.

In the end they simply followed the drops of blood. In one spot the needles and ferns were smeared with blood, as if the animal had stumbled and gotten up again. The fall may have driven the arrow deeper, since they found the pachy about twenty yards farther, dead.

Dr. Shanker gutted and cleaned the carcass.

Hungry as he was, Julian was sickened by the process. Their diet for weeks had been small mammals, fish, and roots; they had not eaten dinosaur flesh since the very beginning. As Julian received the thick, bloody chunks that Dr. Shanker carved from the thigh, the smell and consistency revolted him. The meat was a dark, almost purple red, with an unexpected yielding quality, as if it would squash to gel in his hands. The thick nobbled skin, mottled green and brown with yellow streaks here and there, reminded him of poisonous lizards he'd seen in zoos. Just under the skin was a whitish membrane layer that slipped and glistened and proved difficult to slice. The meat looked, felt, and smelled utterly foreign.

However, there was nothing to do but eat it. Yariko was working on a fire a short distance away, her back carefully turned to the dead animal.

Julian handed her a drooping, heavy mass impaled on an arrow. "Your dinner, my dear," he said, making an effort.

"Why thank you," Yariko said. "It looks wonderful. Did you make it yourself?"

They bantered as they toasted bits of meat, and Julian felt better when his chunk began to look more like food as it cooked.

"What do you think?" he asked Yariko, as he studied his portion, turning it this way and that to see if it was really cooked.

But Yariko was looking over his shoulder with a strange expression on her face. "I'm going to be sick," she said.

Julian turned in time to see Dr. Shanker pop a tiny, dripping piece of raw flesh in his mouth.

"More moisture," Dr. Shanker said, realizing he was being stared at. "You're making a mistake, cooking all the liquid out of it. We have no water without the river."

Julian and Yariko turned their backs to Shanker and Hilda, and tried to enjoy their own meal.

As they finished it began to rain heavily, as if to purposefully contradict Dr. Shanker. Long strings of water cascaded from the treetops and drummed on the ground, washing away the needles, churning the forest floor into mud. They tilted back their heads and drank, and cupped their hands for Hilda to drink. The fire was flattened into soggy charcoal and the blood washed from the carcass.

It was a shame to leave the rest of the animal when they'd hardly taken any meat off, but not even Dr. Shanker was willing to carry raw flesh through the forest. They each saved a well-cooked piece that they wrapped in a leaf and carried on a stick. Dr. Shanker picked through the small mound of viscera he'd thrown to the side and cut the bladder free. He washed it out in the rain, filled it with water, and tied off the top. Julian grinned, watching him: little did Shanker know it was of immense scientific interest that a pachycephalosaur even had a bladder.

The rain slowed and then stopped altogether by midday.

The afternoon was sunnier, and Julian was able to figure out the points of the compass. However, although they walked roughly northwest, the river obstinately refused to appear. It must have taken a meander, looping farther north. There was nothing to do but keep going, and hope.

Finally they struck a dry streambed worn into the soil, snaking between the trees. That caused excitement; surely it would lead to the main river eventually. Julian could not resist picking up stones and splitting open the shelves of sedimentary rock, looking for fossils. Stream beds were excellent places to look for fossils, with the types found dependant on the layers of rock exposed.

In the twentieth century, the geographic region of the Hell Creek Formation was particularly rich in Late Cretaceous fossils; but now in the Late Cretaceous itself, the stream bed was filled with fossils from a much earlier date. Julian found impressions of half-inch shells, sometimes mingled and crammed together in tessellated patterns as if hundreds of the mollusks had died and settled on top of each other.

These fossils dated from the Cambrian, several hundred million years earlier, living and dying in such a distant past that by comparison dinosaurs were modern animals, indistinguishable from humans. If he and the others had been thrown back to that era in time, the earth would have been utterly alien, unrecognizable, although probably much safer.

The terrain began to change again. The giant *Metasequoia* thinned and the gaps were filled more and more with smaller conifers. The undergrowth became tangled again and difficult to walk through, and the ground was softer. They were nearing the river at last.

Hilda had completely recovered her good humor and the swelling around her eye had gone down. At the same time she regained her enthusiasm for chasing after animals, and several times came near to being kicked again by frightened pachys. Julian wondered how many trials it would take before she learned, and if she would survive until she did.

Once, she dashed off through the bushes, snarling, stopping not far away to bark at some unfortunate animal that she must have cornered. They crashed through the bushes after her, partly to rescue her if she needed it, and partly to see the animal. Dr. Shanker was hoping it would make a good meal, if Hilda brought it down.

They struggled out of the bushes, scratched and itching, beating aside the bracken with their spears, and stepped into the clearing. Julian was scanning the nearest tree, looking for a small mammal, treed and hissing down at Hilda from a branch. But as soon as Yariko struggled free of the bushes she grabbed his arm to stop him and pointed across the clearing.

Julian was so startled that he felt goose bumps pricking on his arms and chest.

Hilda was tormenting a monster.

She was barking in a frenzy from just beyond its reach. Its bulk was partly screened by leaves and by the bare twigs of deciduous bushes. But from the back end of its shell to the front of its snout it must have been nearly thirty feet long, another example of the gigantish trend of the Late Cretaceous.

It had a blunt, triangular head, wide and flat, knobbled and plated on top, small in comparison to the body. It held its head near the ground as if in a defensive posture, or in threat, its beaked mouth gaping and showing a single row of small, foliated, plant-grinding teeth far back in the jaw. The body rose up behind the head, armored with rounded plates of bone, like a monstrous turtle. A few small birds fluttered above it. Its rump was raised but it had not yet turned around to present its club-tipped tail. In fact the creature did not seem very alarmed about the puny, noisy mammal that had disturbed it, and after silently threatening, it turned back to feed.

Julian stood amazed, watching. An enormous, bluish tongue reached out of the mouth and curled around a twig, pulling the leafy end into the animal's beak. Such an organ had been suspected by paleontologists, because the skulls of ankylosauria had well-developed hyoid and entoglossal processes to anchor the tongue muscles. Another guess was that most of the digestion would have taken place inside an extensive gut, since the teeth were too small and weak to process the food effectively.

One symptom not much discussed in the journals was that the fermentation of leaves in the gut produced an incredible smell whenever the animal dropped feces, which it did at that moment, in huge wet quantities.

That was enough to make them all back away. Hilda, however, continued to pester it, darting dangerously close and snarling, but the animal snapped back in a halfhearted way, glancing at her without much concern.

Dr. Shanker strode forward. "Hilda! Get away from that! Come back here!"

He made a grab at the scruff of her neck and caught her after several tries. He slipped a rope over her head and began to haul her away while she strained against it, still barking and snapping fiercely, as though convinced she was larger than the monster.

But the animal seemed to think that Dr. Shanker was joining the attack. It began to turn away from him, lumbering on its ungainly, trunk-like legs, the back ones longer than the front. As it turned, snapping and bending the branches around it, Julian saw clearly the ring of spikes curving out from the edges of its armor. Any carnivore foolish enough to attack this animal or try to turn it over would have a bloody mouth for the trouble, and very possibly a fatal gash.

He thought it was maneuvering to lumber away into the woods; but as soon as its back was turned the tail lashed out with astonishing energy considering how sluggishly the animal had been moving. The tip of the tail was weighted with lumps of bone in the shape of a medieval mace. If Dr. Shanker hadn't ducked in time his skull would have been smashed. He hurried back to Julian and Yariko.

The ankylosaur began to back toward them. Even Hilda decided it was not worth antagonizing further. They quickly scrambled away through the brush, Hilda in front dragging Dr. Shanker by the homemade leash. Fortunately the ankylosaur did not follow them very far; when Julian glanced back he saw it settling down to feed again.

Later the same day they reached Hell Creek. It could be heard plainly from a distance, crashing and rumbling, and when they finally emerged from the forest they stood on a rocky cliff, thirty feet high, looking down at a shallow, fast-flowing stream. The raft would never have made it through this stretch, had they continued on the river; clearly luck was with them still. A mist of droplets hung in the air, and the spray flew up into their faces. It was difficult to talk over the roaring water.

The opposite bank of the river was lower. Open lands stretched out beyond for miles, speckled with scrub and here and there a coniferous tree. The sun had almost set, spreading a dull red light over the sky and the plains. Jutting up against this redness, in silhouette along the western horizon, stood the first low slopes of the infant Rocky Mountains. They were nothing like the lofty peaks Julian had visited in his own time.

He followed their probable path with his eyes: it lay toward the very first of the slopes. But he could not help thinking the chances were still slim. The river now bent south so it would be no more use as a guide. Miles of broken, rocky ground lay before them. It was all too exposed; there were few trees to climb, and even less underbrush. The hills were likely cooler at night, and there'd be a scarcity of food and water. It all seemed impossible, especially within the remaining few weeks. Impossible—but they would keep trying. Julian was not discouraged. Instead, he felt strangely excited. Their goal was suddenly a reality. Finally, it was in sight.

Yariko touched his arm and pointed. One great swath of the plains, beginning near the river but stretching away at least a mile, looked darker than the rest. Julian thought it looked like turned-up earth, lumpy with mounds and boulders, as if some giant had begun to till the field and then found it too rocky to cultivate; but the distance was so great that the boulders must have been enormous up close.

Suddenly he realized that they were not rocks or clumps of earth at all, but animals, bulky *Triceratops*. Like the bison or the wildebeest they were gathered together in uncountable thousands, stretching beyond all imagination, sending up a haze of dust that caught the red light of the sunset.

They watched for a long time. Dr. Shanker put three fingers to his forehead and mimed a lumbering, rhinoceros kind of animal, and Julian nodded vigorously. Dr. Shanker grinned, his whole leathery face bunching up with pleasure. Yariko also grinned; she took Julian's hand and held it tightly, hurting his fingers, but he did not complain. He squeezed back.

Finally they stepped away from the river and made camp under the eaves of the forest. Yariko started a fire and when Hilda suddenly appeared with a raccoon-sized mammal in her mouth, Dr. Shanker confiscated half of it for their dinner.

"And what can we expect from this new terrain?" Yariko asked, as she turned slices of meat on a makeshift grill made of twigs propped on stones.

"Hopefully not lava flows, ash falls, or other unpleasantness," Dr. Shanker put in, also in a loud voice to be heard over the river. He was clearly in a good mood, now that Hilda was back to normal and the river had been found again.

"I don't know about other unpleasantness," Julian said, "but I doubt we see any volcanoes in action, or not what most people call action. Most of the magma activity is below the surface here, and the geology takes place at appropriately geologic time scales." He shifted his position and dug out a small rock, about two inches across, that was too sharp to sit on. He glanced at it before tossing it away. Granite: not the shale or other sedimentary rocks that had so far been common. No fossils in this rock; but it was interesting nonetheless.

"Look at this," he said, passing it to Yariko. "Notice the texture."

She turned it in the firelight and the crystals gleamed faintly. "Texture," she said, thoughtfully. "Well, it's not smooth and flat like the fossil rocks you keep breaking open. It's more grainy. I can see big grains in it, different colors."

"Exactly. It has large crystals, mostly quartz and feldspar. It's primarily made of silica and aluminum; lighter elements."

Yariko handed the rock across the fire to Dr. Shanker.

Julian went back to toasting his chunk of meat. "That may be all the volcanic activity we see."

"This is volcanic?" Dr. Shanker looked doubtful.

Julian nodded. "It was formed in the slow cooling of a magma intrusion deep underground, where temperature and pressure are high. If you want to get technical, volcanism is an extrusive or surface process, and plutonism is the intrusive, or below the surface, equivalent. But they both involve magma."

"Then how did this piece get here, if it was formed under the surface?" Yariko asked.

"Uplift, and then erosion of the surface rock. The river brought it here."

"And is this part of your Boulder Batholith, that we're trying to reach?" Dr. Shanker looked at the stone with greater interest.

"No, I doubt it. Those are still too far away. This little piece probably comes from a much smaller formation closer to the surface. But it's a sign we're entering the right kind of terrain."

Dr. Shanker tossed away the stone along with his meat-toasting stick. "I suppose we'll find plenty more, as we go on," he said. "It seems a hopeful sign. Lecture us more later. Tomorrow. I say we sleep now; we could all use a rest." He yawned noisily, and Hilda yawned too, with an almost exact imitation of his facial expression. Julian and Yariko both grinned.

The trees by the river were short and none of them were good for climbing. Dr. Shanker curled up on the ground with Hilda, and by habit, although there was no reason, Yariko and Julian found a spot about thirty feet away and out of sight. They crawled under a bush, into a soft pile of leaves.

The pounding of the rapids deadened all sounds from the forest; even Dr. Shanker's snoring went unheard. The background rumble gave Julian a sense of seclusion and security that he hadn't felt in a long time.

"It seems hopeful, doesn't it?" Yariko said out of the darkness beside him, her voice so close that he could feel her breath on his cheek and detect, faintly, the sweet smell of the berries they'd eaten as a desert. "Reaching our goal, I mean. We might really make it, even in the few weeks left. Tonight was the first time I felt confident."

"Yes . . . we do seem to have made remarkable progress," Julian

151

answered slowly. He too felt more hope, even certainty of their chances, than ever before. But he felt something else too. With the new hope a sense of dread had begun to creep in: not dread that they'd be stuck in the Cretaceous forever, but dread that they would leave it. "We haven't talked much about how we'll live if we don't get back," he went on. "Back to our original time, I mean. After all, even if we make it a thousand miles, we have only a vague idea of the right spot; and we can't know if the lab will be functioning."

"I suppose it would be practical to have a plan," Yariko replied. "But the goal is to not stay here, so that's what we've been focused on. Isn't it?"

Julian caught a slight hesitation before her last two words. "You said you were getting used to living like this, in this time." He had been thinking about that since she'd said it, in fact. "Maybe I'm getting used to it here too, or the daily requirements have made life more present, more immediate, or something. It's been easy to forget any other world ever existed. Not always, of course, but more and more often I forget, for long stretches at a time."

"I do, too," Yariko said, and she sounded sad. "But I don't want to. It's just hard to compare."

"Yes . . . it's hard to think about our past lives when we're so busy in this life. I know our goal is to revert back to the lab, but sometimes it seems like that's not what we're focused on."

"I don't understand," Yariko said.

"I mean living—just living, day to day, being with each other, finding food and shelter and keeping ourselves safe in the wild: isn't that what we're really doing?"

Yariko was silent for so long that Julian wondered if she'd dropped off to sleep. When she spoke it was in an even softer voice than before, and he had to strain to hear her. "And in a sense it's a better way to live. No external pressure, everything depends on us, on our own strength and skill and, well, there's no one from out-side of ourselves expecting us to live a certain way or meet certain goals. Certain impossible goals. So many goals that have no hold on my real self, but only on the self I think I have to be. . . ."

"Like Dr. Miyakara?"

"Like Dr., yes, and other things—other titles."

Like Mrs. Somebody, Julian thought suddenly, but he didn't say it.

"And what about you?" Yariko's voice was brighter now but there was still a tug of sadness behind it.

"You mean, is this life better in some ways?" Julian shifted his position and put his arm over Yariko. "Well, I miss things, of course, like my lab and all the work I was doing, and some people, and, well, indoor bathrooms with hot water, to be honest." Yariko snorted, and he went on. "But I like it here too. Because how can a fossil compare to the real thing? And because . . . because of present company." He stopped in embarrassment.

"If we ever returned," Yariko said, "you wouldn't lose my company. Not now. Not knowing what I know, about you . . . and myself."

Julian was quiet a moment. "Sometimes you joke about starting a village," he began, and then paused, changing what he was going to say. "The fact is, I would be happy in the Maastrichtian with you. I would be happy at the University with you. Wherever. It doesn't matter."

He fell silent. He felt foolish, especially as Yariko said nothing for a while. He could hear the quickened sound of her breathing.

Then suddenly her head was on his shoulder, heavy and comfortable, and she gave a little laugh. "No one's ever said that to me before: 'I'd be happy in the Maastrichtian with you.' No one's ever felt that way about me."

Julian's heart tightened and then seemed to expand with relief. "Takes a paleontologist, I guess," he murmured, but he was already on to more serious things, his hand stroking her hair and then moving down her back to pull her in closer. When he found her lips her arms went around him and her whole body shaped itself against his, and it felt absolutely right.

In the morning they stood at the cliff again and looked across the river. The southeast side was thickly forested, but the northwestern side was open, vast in the bright sun, brown with brush now losing its leaves, dotted with occasional trees. The *Triceratops* were gone but their trail was clear, a brown gash meandering across the plains. Julian wondered if they migrated south for the winter, mild though the seasons were. The trail, having fetched up against the river, turned toward the west and followed the bank for about a mile. There it disappeared from view. The creatures had probably found a shallow spot, crossed the river and then continued southward into the plains.

153

As clear as the trail was the smell. Julian had been close to herds of buffalo and was reminded of the sharp, musky odor that seemed to emanate from the massive bodies. But now the odor emanated from the empty air and trampled ground, and it was a hundred times stronger, stringent and overpowering.

The going looked easier on the opposite bank: more open, less steep and tangled. Julian suggested they cross and follow the *Triceratops* trail as long as it went west, it being a clear and easy path. The truth was he wanted to travel in the wake of those immense and wonderful animals.

Hilda shot down the rocky, uneven cliff face with no difficulty and stood waiting at the bottom, wagging her tail and laughing up at them. The humans followed more slowly, mainly because Julian helped Yariko rather more than was necessary; she would have done better alone but she accepted the hand he lent through the steepest parts. At the bottom, the two stood together on a slab of rock, his arm over her shoulders, and blinked and squinted in the frigid spray.

Dr. Shanker stood beside them with Hilda. He had a wide grin on his face as he looked at the water. Julian, happy himself on what seemed the loveliest morning he'd ever seen, didn't wonder at the man's delight.

"Can we cross here?" he shouted over the roar of the water.

Dr. Shanker turned and looked at him, still grinning. "You might have to go single file," he yelled back, and gave them an exaggerated wink.

Julian felt Yariko stiffen under his arm; but before he could retort, Dr. Shanker spoke again.

"Come on, Whitney, don't scowl. It's about time, really. I didn't think even you could be that shy, and I know Yorko isn't. To think of all those days on the riverbank that I confined myself to the bottoms of trees, risking the predators. . . ." Hilda laughed up at him, wagging her tail as she caught the general mood of excitement, and Dr. Shanker reached down to ruffle her ears. "And I get stuck with the dog!"

Julian couldn't help smiling even as he felt himself blush; in fact his grin rivaled Dr. Shanker's. Yariko pushed his arm away.

"If we're going to cross, we'd better have a look—single file," she said.

Julian reluctantly looked away from her to study the river. The rapids were a daunting sight, pounding and tearing through jagged boulders. But the river was only thirty yards wide now and scattered with so many large rocks that it didn't look too hard to cross, as long as they didn't lose their footing. In fact, it almost looked as though a line of rocks had been placed by plan, stretching from one bank to the other.

"Well well, someone's left a bridge for us," Dr. Shanker said, indicating the path with his spear. Hilda leapt joyfully through the spray from stone to stone and was soon on the other side, where she shook herself dry and gave an excited bark.

Dr. Shanker followed, using his spear to balance as though he were in a circus on the high wire. He paused before the last leap, a six-foot gap between the last boulder and the shore. Then, gathering himself, he sprang across and landed squarely on the rocky bank. There he stood, leaning on his spear, smiling back as if pleased with himself for such a fine athletic display.

Yariko came next, and Julian brought up the rear. The stones were more slippery than he had expected. He tried to use the spear as a staff, prodding the point of it into the foamy water, searching for the bottom, but the current nearly yanked the pole out of his hand. The roar of the water was almost painful as it boomed off the rocky cliff. It was disorienting, and he felt dizzy. At each jump from one stone to the next he expected to slide into the freezing river and never come out again.

He had just reached the last stone and stood precariously on the uneven surface, measuring the distance to the bank, when he happened to glance up. Looming over the bushes, staring down at Dr. Shanker and Yariko, was the enormous slate-gray face of *Triceratops horridus.*

Julian froze, the back of his neck tingling. Hilda was oblivious, probably overwhelmed by the enormous stench of *Triceratops* all around. She wagged her tail and cocked her head at him. Yariko mouthed the word "Coward!" evidently thinking he was afraid to make the last jump.

At that moment the creature barked. It had a rasping tone, unpleasant and much higher pitched than expected given the animal's size, but quite loud nonetheless. The sound, rising above the thundering of the river, had an instantaneous effect. Yariko, Dr. Shanker, and Hilda turned to look behind them; and Julian lost his footing and tumbled into the river.

Part III

The Triceratops Plain

The extinction of a species has been popularized by images of lonely, one-of-a-kind animals lying down to die in despair on dusty plains. In reality, the poignant last-of-its-kind scenario is an invention of modern man, who has also, unfortunately, made it an actuality in some cases. But through most of Earth's inhabited history, species have come and gone through undramatic and almost imperceptible processes: mingling with others, change of habitat, genetic modifications, gradual loss of diversity and fecundity.
—*Julian Whitney*, Lectures on Cretaceous Ecology

– THIRTEEN –

The current took Julian before he could cry out. He struggled to keep his head above the surface; but with all the foam around the rocks there was no surface to speak of. He caught a mouthful of water and choked, then smashed the side of his head against a stone. His sneakers touched the pebbly bottom for an instant. But he was carried away again, turning and scraping, eyes clenched shut against the spray. Through it all he clutched the spear, and it clunked and bounced against the rocks around him.

When he finally opened his eyes he was below the falls and in the center of the stream, momentarily safe from the rocks, but moving rapidly toward a sharp bend in the river. He tried to swim toward the shore but was so chilled that his arms would hardly move. He gave up and simply drifted, still clutching the spear but not even remembering it. He felt quite peaceful, as if in a dream.

He had no sense of distances. One moment he was drifting down the center of the stream and the next moment the current slammed him against the rocks at the bend. Everything turned into foam again, noise and spray, but he had the presence of mind to clutch onto a point of rock. For a while he hung there, the force of the water pressing him against the rough surface.

The whole adventure had been so rapid and unexpected that Julian's mind had not caught up with it yet. But as he clung to the rock he realized for the first time how close to death he was. No dinosaur could have done him in more effectively than the freezing water. A shaking, panicking terror came over him. He struggled, and with the help of the spear managed to scramble onto the rocks. That was as far as numbed muscles could go. He would have done better to crawl out of range of the spray, but he lay quietly, one foot still dangling in the river, and the world went dark for a moment.

• • • • •

Yariko looked straight into the ten-foot-long, armored, and horned face of *Triceratops horridus*. It was so close that its horns almost touched her. Never had she imagined anything so big. It loomed over her, filling all her vision and blocking the sky and the trees. She was unable to move. She stood with her mouth open and her eyes trapped as the scent surrounded her like a cloud.

Close as they were, a myriad of details appeared that she'd never imagined. The face was not gray: it was rust black like the mold on the north side of a tree, and the frill and horns were a light brown unevenly splotched with white. The eyes were small, or appeared so in the vast face, and the corners seethed with flies. But what caught her attention the most, ludicrously, were the ears. Even in her terror it amazed her that the thing had visible ears: long creased ears like a rabbit's, leaf shaped and covered with hair-like bristles, sticking straight out sideways from behind the eyes. Had any paleontologist guessed that *Triceratops* had such external ears? She remembered Julian talking about ear holes just behind the eyes, but she thought that even he would be surprised at the reality.

Hilda startled her out of her moment of shock. With a woof that rivaled the *Triceratops'* bark she shot around Yariko, nearly knocking her over, and took up a protective stance between the humans and the dinosaur. Yariko watched in amazement as this puny mammal, without horns or frill, faced down a monster whose head alone was ten times her size.

Dr. Shanker was no less overawed. He had turned toward the animal and then staggered backward, falling over his spear and rolling nearly into the river. Now he picked himself up rushed forward to grab Hilda. "Leave it!" he shouted, but his voice was a tiny squeak lost in the roar of the river and the sharp, rasping bark of the animal.

He grabbed Yariko's arm. "Back away slowly," he hissed in her ear. "Don't look it in the eyes."

Yariko gave a snort on the edge of hysteria. From their six-foot proximity it was physically impossible to look the thing in both eyes at once.

In answer, the *Triceratops* lowered its head further and shook its horns at them; horns a yard long and very sharp at the tips.

"Julian," Yariko whispered as they took a step back, and another. "Did he make it back across?" She was sure the thing wouldn't follow them into the river; then again, she wasn't sure at all.

They took another step backward. Hilda didn't move; with one front paw raised and her lips drawn back from her white teeth, she kept up a continuous, deep growl that vibrated in Yariko's head and chest.

"It'll kill her," Yariko said.

"She's giving us a chance." Dr. Shanker took one more step back; now they were at the edge of the water.

"Julian!" Yariko called. She looked quickly behind her, into the river. Only seconds ago she had been laughing at his hesitation to jump; now he was gone. The river tumbled on with white spray over boulders, but Julian was gone.

"He must have crossed back!" Yariko shouted. She felt confused by the rushing water beside her feet.

"I don't see him on the other side," Dr. Shanker yelled back. "Anyway he couldn't have crossed that quickly."

Then Julian was forgotten again as the *Triceratops* gave a sort of bellowing roar, rough as gravel and ending on a high note that hurt the ears. It shook its head again and the wall of bushes shook with it, making a clacking, twiggy sound. Then Yariko saw another head poking its way through the bushes, lower down. It was identical to the first but in miniature, being no longer than Hilda from parrot beak to scalloped frill.

"Wonderful," Dr. Shanker said. "A mother protecting its young."

Yariko only caught a word or two but she didn't care. "The water!" she yelled. "Maybe he fell in the water!" She swung around to stare downstream. There was nothing, no dark head bobbing in the white swirls.

"If he did, that's the end of him," Dr. Shanker shouted back.

Together, they began to inch their way downstream, expecting that at any instant the creature would charge and drive them into the river; or worse, spear them with those horns.

"Hilda! Come on, girl!" Dr. Shanker whistled as well but Hilda remained firmly in place.

The *Triceratops* did not. The monstrous head was pulled back through the bushes only to reappear downstream, closer to the river and only thirty feet away. Now they could see the whole animal: it was at least ten feet tall at its arched back and perhaps thirty feet long, bulky, stocky, and unmovable. It was blocking their path.

Dr. Shanker still held Yariko's arm. Now he pulled her in the other direction. "This way!" he shouted. "Don't move toward it!"

But Yariko had seen something. Just downstream, fetched up against the stony shore and slamming into a rock over and over again as the water went by, was Julian's bow.

She pulled her arm from Dr. Shanker's grip. "Look! He's been washed away. He'll drown, or get battered to death. We've got to find him. That way!" Now she grabbed his arm and began to pull.

He resisted. "We can't go that way. We can't get past that thing until it decides to leave. Whitney's on his own until then."

"We haven't even tried," Yariko shouted. "And there's no time to wait." Before Dr. Shanker could react, she yanked the spear from his hand, and screaming, "Get out! Go on! Get!" she ran straight at the *Triceratops*.

● ● ● ● ●

Julian came to shivering uncontrollably. With a great effort of will he dragged himself farther up the bank, into a bed of ferns. There was nothing to do but wait for the others. Fortunately, he had fetched up on the western side, and it couldn't be that far from where they'd crossed. Surely Yariko and Dr. Shanker would be along within the hour. He refused to let in the thought that he'd lost her, really lost her, just when they'd become close.

After a while he struggled out of his wet clothes and hung them on the twigs of a nearby bush. The battered sneakers were stuck to his feet like suckers and he barely had the energy to fight with them; but he managed in the end, mainly because he was desperate to get his soggy jeans off. Then he sat down, hugging his knees, trembling, waiting for the warmth of the sun to take effect. As the numbing cold disappeared various bruises began to hurt. His head ached from being smashed against the rocks.

After what felt like hours there was still no sign of the others. Julian struggled to his feet, shaking, and dressed. The T-shirt had dried well enough; the fabric was so worn and thin that it could not hold water very well. But the jeans were damp and uncomfortable. He picked up the spear, and leaning on it heavily at every step, still weak and trembling, followed the river upstream. The ground was stony and choked up with bushes. On every rock he left behind a wet splotchy footprint.

Finally he reached the place where they'd crossed. There was no one in sight.

It was as though the *Triceratops* had never been there, or his companions either, for that matter. The ground showed nothing even to Julian's footprint-trained eyes. He stumbled up a slope of stones and sand that lay between the river and the thick, rising wall of bushes. The effort of forcing his way through was almost too much, but finally he managed, scratched and weak, to push his way out the other side.

He stood in the trampled path of the *Triceratops*.

It was like a river of its own, a hundred yards wide, wider than the actual river in fact, stretching between green and brown banks of tall bushes. The ground was all mud and rock, tangled up with the flattened remains of scrub. The animals seemed to have eaten all the leaves and smaller twigs and then trampled the fibrous trunks into the ground. But the path was empty. As far as he could see in both directions, nothing moved except a few small birds, the size of sparrows, pecking at insects in the mud.

The *Triceratops* had vanished, and so had Julian's companions: Yariko, Dr. Shanker, and Hilda. The sound of the river behind him seemed to accentuate the dreadful solitude.

He crawled back through the bushes and sat on the stones at the river's edge, trying to think. His head was throbbing so much that he could hardly sit upright. But many times before, in the grip of a bad flu, he had staggered to the office and worked on a project that needed to be finished. He called on that same kind of energy now, despite the rising fear and desperation over being alone.

Forcing his mind into action, he made a list in his head of all the possibilities, and for each one he drew a mark on a flat stone, scraping it with the pointed edge of a pebble.

First: Yariko and Dr. Shanker had been searching for him, but somehow they'd missed each other. That was hard to believe, since after crawling out of the river he had collapsed within plain sight of either bank. Dangerously exposed, in fact, now that he thought about it.

Second: they'd been killed, and the bodies had somehow been removed—eaten by a passing predator, perhaps. At that thought panic almost overwhelmed him; he felt like jumping up and rushing wildly around, calling for Yariko. But then reason took over. There was no blood anywhere, no sign of a struggle, and anyhow it was hard to believe they'd all been killed outright, including Hilda.

Third: they had escaped the *Triceratops* by scrambling back across the river. He peered across, shading his eyes against the late-morning sun, seeing nothing over the top of the cliff except the blue of the sky behind it.

The first possibility seemed the most likely. They might search for a time, and not finding him near the river they might make for the *Triceratops* trail that had been their goal for the day. It was certainly more open than the tangled brush near the river.

Julian knew that if he didn't find them soon he should continue the journey toward the west and try to catch up with them, if there was anybody to catch up with. If there wasn't, then he might as well be traveling west as any other direction. But he did not have the energy to get up at that moment. He crawled into the bushes, curled up in a ball, and tried to sleep.

For a long time all he could think about was Yariko, the dearest person to him in sixty-five million years, possibly dead, maybe in danger somewhere; possibly trudging along with Dr. Shanker on a hopeless search, convinced that he himself was dead.

If he had been able to think clearly, he would have known that his companions hadn't gone far. He would also have known that *T. horridus* was not an easy animal to get away from, especially once enraged.

• • • • •

The *Triceratops* lowered its head, presenting horns and bone frill to its diminutive attacker. Perhaps it thought Yariko was one of the small bipedal carnivores, maybe a type of raptor. Even if she were, she wouldn't have much chance against the monstrous creature's defenses, not to mention its bulk.

Dr. Shanker watched half in frozen horror, half in disbelief. He was unable to move or even to yell. Yariko seemed almost to run up against the deadly horns before stopping. She was like the mouse standing up to the lion; her arms waved wildly over her head, the spear flailing as she jumped up and down in place; and she yelled without pause, a string of words repeated over and over: "GO AWAY! LET ME BY! GET OUT! GET OUT!"

Then Hilda reappeared, leaping out of the bushes to join Yariko. She did not take up the protective stance this time; she seemed to understand that Yariko was now the aggressor, and she joined in the attack as an equal.

The *Triceratops* pulled back and shook its head. It gave another rasping bark, which turned into a roar like a bull's, only ten times louder; then it leaped forward.

Yariko turned with a shriek and pelted back along the bank.

Dr. Shanker expected her to stop where he was, but she flew right past him, the butt end of the spear catching him on the thigh with a stinging blow. "Run, you idiot!" she screamed, and was gone.

He looked up and realized he was between the *Triceratops* and Yariko. And the *Triceratops* was moving faster. For an instant he saw its nostrils flaring red as blood and heard its snorting breath; then he too turned and ran, with no thought in his head except panic, and the snorting monster at his back. Somewhere in the back of his mind he was aware of Hilda's barking, but it hardly registered.

Yariko was far ahead now running upstream. Dr. Shanker stumbled on the rough, stony ground, trying desperately to keep his balance and to catch up; then Yariko too stumbled, and fell to her knees.

"Into the river, quick," Dr. Shanker said, rushing up and grabbing her arm. "We can't run forever." He jumped for the first large boulder in the frothy water, and then the second. Yariko had no time to think; she followed without looking back. Indeed, there was no need to look when she could hear the immense breathing and grunting almost at her back.

Halfway across the river, they turned to look at the shore. Hilda was barking now, standing on the rock nearest the bank. Their attacker seemed unwilling to enter the river. Perhaps its young wasn't able to cross there.

"That was brilliant," Dr. Shanker said, scathingly. "Enrage something that big, with horns like that. Great move. Now we're stuck in the middle of the river, way upstream too."

Yariko didn't answer. She began to pick her way downstream again, slowly, from rock to rock. Very quickly she came to a spot where there was nothing but water ahead; a large pool, with no protruding boulders for at least thirty feet.

"You won't catch up with Whitney that way," Dr. Shanker said from directly behind her. "And don't even think about jumping in the water. You'll drown."

Yariko turned on her rock to face him. Hilda was standing on the same rock, all four feet bunched up under her, looking unbalanced and unhappy. "He may be drowned by now. We have to get downstream." She was panting, sweaty, and still had a wild look in her eyes.

"Follow me," Dr. Shanker said. There were enough rocks to take them back to the bank, behind the *Triceratops*. "If we're quiet, maybe it won't notice."

Hilda made the decision for them by leading the way; she'd had enough of teetering on rocks. Yariko followed, then Dr. Shanker. There was no need for silence as the water covered any noise they made.

Yariko watched Hilda sway on a small rock and then launch herself at the bank, landing with her front feet in gravel. She dragged herself out and shook the water off her fur.

"Go on," Dr. Shanker said, impatiently. "I can't balance on this rock forever."

Yariko judged the next leap, a long one, but before she could jump she felt her foot slide on the wet rock and go into the water.

It was terribly cold. She thrashed out wildly to regain her balance, and managed to whack Dr. Shanker on the jaw. He grabbed her around the middle and deposited her solidly back on the rock.

"I didn't mean go in," he said, still gripping her wrist.

"Cold," she said, her teeth chattering at the mere thought of being submerged. "Too cold. If he stayed in long. . . ."

They made it back to the western shore. There was no sign of the *Triceratops*.

• • • • •

Late in the day, Julian woke feeling warmer but so achy he could hardly move. As soon as he sat up the fact of his aloneness rushed in and almost overwhelmed him. He closed his eyes and pictured himself leaping up and joyfully dashing toward the sound of Hilda's bark or Yariko calling his name. He pictured it so vividly, and wished for it so hard, that a rustle in the bushes set his heart beating in wild hope. But nothing emerged. Nothing moved, anywhere. He was alone.

It was worse than fighting the dromaeosaurs. It was worse than the worst night on the river, with thoughts of Frank as they'd left him, the stink of the rotting jungle, hunger, despair; it was worse than anything he'd ever known.

What if one of us is left alone? Yariko had asked. Forever?

What if he was the only one alive: the last person on earth, for the next sixty-five million years?

If he came to believe that, everything was over. With sudden determination, Julian picked up his spear and set out along the riverbank, upstream, in search of his companions.

After a mile, he reached the point where the *Triceratops* had forded the stream. Their path turned sharply, plunging across the stony river bed and disappearing into the plain to his left. The cliff on the opposite bank of the river was lower there, eroded, now a well-trampled path in dark soil. The herd had crossed the river and gone on.

But one of them had not succeeded. It lay dead in the river, quite still, blending in against the gray stones so well that Julian hadn't even noticed it at first. It was a small *Triceratops*, maybe a youngster.

Here was enough meat for his dinner and for a hundred other people too.

Julian had only just escaped the river; but with a grin and a shrug reminiscent of Dr. Shanker he clambered out over the rocks. The water snarled below his feet. The animal was near the middle of the stream, and he climbed on top of it and squatted, resting for a moment. He felt like a caveman on a mammoth, he and his spear: a very smelly mammoth, at that. And he felt like laughing, or crying, or maybe both. He stood up, balancing carefully on the hard nubbled hide of the animal, to spy out the area from this convenient high point.

The sun was almost down. Over the low cliff to his left the forest grew thick and was already turning black in the dusk. On the opposite bank, to Julian's right, the bushes stretched out across the stony plain, blending into a colorless smear in the twilight. Only the river was still bright. Nothing moved except the sparkling current.

Julian fetched out his knife, tangled into the frayed cloth of his jeans. It was unusually stout for a pocketknife but wasn't designed for cutting through *Triceratops* hide. It was also rusty. He tried to clean and sharpen it against a rough stone but the rust remained. Clambering across the animal, he began cutting at one of the hind legs, which lay partly submerged.

The skin was incredibly tough, leathery and almost slippery, which was surprising: he'd expected it to be dry and rough like rock. It took some time to slice through the stuff. Now and then as he worked Julian looked up and stared around anxiously. The Maastrichtian world must have had its scavengers. Some people believed that *Quetzalcoatlus*, that forty-foot airplane of a flying lizard, must have been a Cretaceous turkey vulture, spotting carrion as it wheeled and glided over the open landscape. But he didn't see anything so large on wings; in any case, it was thought to be a more coastal animal.

The hide, once sliced through, proved equally tough to pull away from the meat. Julian had to brace his foot against the animal's thigh and tug with both hands. When he had hacked and torn away a piece several feet square he washed the blood from the skin and laid it on a nearby boulder. Then he hacked at the thigh muscle. Surprisingly little blood ran from it, but it reeked nonetheless: of musk, blood, and death. Julian gagged and then had to stop and put his T-shirt over his face for a moment.

Within minutes, despite the slight remaining warmth of the flesh, his hands were nearly frozen and the knife almost dropped into the river. But at last he had a good pile of meat stacked on the hide, enough for perhaps ten days. He wrapped it up in the hide and scrambled back across the stones to the riverbank.

There several problems presented themselves.

The meat was heavy. It also smelled strongly and left a trail of red drops. He sat down on the ground and thought. Clearly the meat had to be dried: removing the liquid would fix both the weight and scent problems. And smoked meat would keep for several days, even if it would be dry eating.

He managed to strike a spark between the file in his knife packet and the bit of flint he'd been carrying around. As the strips of meat began to smoke, laid out on flat rocks, the aroma seemed to drift up around him like a beacon to all tyrannosaurids within five miles.

For a long time Julian sat beside the fire, adding twigs and leaves, turning the bits of meat. He knew almost nothing about tanning a hide, except that it was important to clean off the wet underside to prevent decomposition. As he sat beside the fire he scraped at the raw bloody surface of the skin with a flat bit of stone, until his fingers were cold and sore.

If Yariko and Dr. Shanker were nearby they must notice the fire. There was no better place to wait for them than at the edge of the *Triceratops* trail. And if they weren't nearby, if they weren't anywhere anymore . . . how could one comprehend being the only human animal to exist in this vast tangle of life? If he contemplated it even a moment longer, if he strained his mind to grasp such a finality, Julian thought he might run screaming into the river again, this time to stay. He crept closer to the fire, taking in its warmth and movement as if it were a sentient thing.

The moon went down and the land around him turned black, accentuating his solitude. The river rumbled beside him unseen. Only the stars were visible, and in the firelight a dim circle of stony ground, a ragged piece of hide, and a prehistoric man squatting by the fire, muddy and unkempt.

Carnivores have evolved several predatory repertoires. "Grapple and slash" predators, typified by modern felines, are ambushers using short bursts of speed to bring down prey. Curved, flattened claws are the main weapons, including large hind claws used for disemboweling. The Cretaceous dromaeosaurids, including Troodon *and* Velociraptor, *were probably such hunters. "Pursuit and bite" predators, like the modern canids, are chasers who use their strong jaws and teeth to dispatch prey. The fearsome tyrannosaurids were likely classic "pursuit and bite" predators. They had tremendously powerful jaws and necks, short but strong forelimbs, and killed by tearing out huge chunks of flesh.*
—*Julian Whitney*, Lectures on Cretaceous Ecology

– FOURTEEN –

1 September
8:10 PM Local Time

"Now that was one weird conversation," Hann commented as he walked back to the station with Earles. "Are you heading home?"

"No. I'm staying until I get more information. That pack of government regulators, that OSHA, might shut the whole thing down tomorrow, and then we'll never know what happened." Earles had been moving at her usual long stride, but now she stopped and turned to face Hann, so that he nearly bumped into her. It was getting dark and a few fireflies blinked near the hedges. A peaceful September evening . . . except that it wasn't. "I'd like to talk to you about Frank," she said.

Hann looked down at his feet while he took in her words. "OK. Shoot," he said after a moment. He took out a cigarette and his lighter.

"Was—is there anything about him that could explain his disappearance? Could he reasonably have walked off and gone somewhere else, without telling family or friends?"

Hann snapped the lighter shut and put it back in his pocket. The cigarette didn't help after all but he drew in the smoke anyway. "No. Not reasonably. I don't know him as well as I could of course,

170

since mostly we didn't grow up together. But when he decided to move here last year, so we could get to know each other, he didn't ask for my help. He was independent, and responsible."

Earles nodded. "That was my impression too. Former marine, wasn't he? Used to discipline. No," she went on, starting to walk again, "I agree. If he went somewhere, it was involuntarily."

"What do you mean involuntary? They took him somewhere by force?" Hann's face took on an angry flush, and for the first time Earles saw a physical resemblance to the guard Frank.

"Perhaps I do mean that," she said thoughtfully, more to herself than to Hann. "So they weren't vaporized, because there are no . . . residuals. And that wouldn't create a vacuum anyway." Somehow, Earles couldn't get her mind off the conviction that those people were inside the vault . . . and then they weren't.

She stopped walking again, abruptly, and this time Hann did collide with her. "Uh, sorry," he mumbled, looking embarrassed and rubbing his shoulder where it had struck hers. "Are you. . . ?"

"Beetles," Earles said. "Beetles and stones. Twigs and beetles. From where, Charlie? From where, and how?"

Hann looked blank; it was a look that Earles was used to by now. He generally trusted her, but clearly he didn't share her fascination with oddball things like twigs and beetles, things that could be picked up outside of any campus door. She wasn't sure if her own interest was worth anything, for that matter, but it was growing stronger each minute.

If such things, living things too, could come from somewhere and appear inside a sealed vault, wouldn't that increase the air pressure in a sealed chamber? Even if they were very small things?

Without warning, Earles turned and strode back toward campus. "You go take a break," she said over her shoulder. "I'll let you know when I need you."

"But where are you going?" he cried after her, feeling a little bereft just when he was opening up about Frank.

"To the lab. To find that mop-headed graduate student. He knows more than he's saying."

• • • • •

It was late afternoon. Yariko's chill had long since been replaced by sweat in the full heat of the open plain. They had followed the river for hours, moving in a pattern from the shore up to the plain and back, trying to cover all possible ground. They had found nothing.

"He must have gotten out before here," Yariko said, as they stood on the bank once more looking at a log jam in the water. "There's no way he'd be washed over this."

It was a thorough jam, with a drop to a calm pool on the other side. There was no body circling in the eddy, no Whitney on this or the far shore, no footprints or any other sign he'd been there. "OK, you may be right," Dr. Shanker conceded. "But then we should be ahead of him, which means we'd have seen where he got out."

"Not if it was rocky. Not if he got there hours ago. He was probably in the river only a few minutes." She was silent a moment, pondering. "I think he got out and made for the Triceratops trail. That's where we were headed, after all. I think we should go back, to where that dead one was in the river."

Dr. Shanker acquiesced, with some misgivings, and they turned and made their way back along the shore. They'd passed the dead animal in the river about a quarter mile upstream. It was clearly the crossing point of the herd. He'd gone out to look at the animal and had seen the sliced-up thigh, the pecked-out eyes, the open belly with pale intestines trailing in the water. He hadn't said anything to Yariko, and he didn't tell her now either. He realized that something as small as a human body could be scavenged to nothing in a matter of hours.

But another search of the area revealed something new: the remains of a small fire.

Yariko was ecstatic. "I knew he was here," she said. "How did we miss this before? And look—he went through the bushes here. See where they've been trampled? It looks like a path."

"Anything could have done that," Dr. Shanker grumbled. "But all right, maybe it was him. Certainly he made the fire." He was relieved, yes, but he didn't entirely share Yariko's elation. "You realize," he went on, "we've seen this now only by chance. Any signs of him will be tiny; and we're searching a huge area. We've probably missed all kinds of signs. And he may be doing the same. We could be here for days."

Yariko nodded. "We've got to think hard about where he is and what he's doing. I say we follow this path."

They emerged from the tunnel-like opening in the brown-leafed shrubs and looked once again on the open plain. The *Triceratops* trail followed the direction of the river and was plain to see for some distance. Nothing moved on it.

Julian's path, if that's what it was, seemed to continue straight toward the west. It wound among patches of low bushes and the occasional gray boulder, a narrow, six-inch ribbon of hard-beaten earth. The lack of twigs, leaves, or even small stones indicated a frequency of use.

"Whitney didn't make this trail," Dr. Shanker said. "In our own time I'd say it was a deer path. Must be some Cretaceous equivalent of deer."

"He didn't make it, but he followed it," Yariko said. She stooped and lifted a stick, broken off at one end and whittled to a point at the other. "This was done with a knife. He was making a spear."

Further on they came across unequivocal evidence: the outline of a human shoe neatly centered within the massive footprint of *Triceratops*.

"I believe we may actually find him," Dr. Shanker said. "It almost seems possible." Then he pointed over the bushes. "Look," he said. "Another dead one."

Away to their right lay the body of a *Triceratops*, an adult. They didn't get too close. The carcass was torn in half and huge chunks were missing from its back end. Looking at the torn-up earth and a trail through the dust and twigs, they decided the animal had broken away from the herd and been chased down. Its side was scored by a series of long, bloody grooves. The whole thing reeked of almost-fresh blood and *Triceratops* stench, and was surrounded by a cloud of flies. It was not that old.

They silently moved back to the "deer path" and hurried away. After a moment Yariko said, "It must have been attacked by several things. All those claw marks on its side. . . ."

"Claw marks, or teeth marks?" Dr. Shanker walked ahead now, clutching his spear again. "Of course, it'd have to be something with an enormous mouth. I wonder what could bring down a Triceratops. There doesn't seem to be anything big enough, from what we've seen."

Had Julian been there, he could have told them what might run down and kill an adult *Triceratops*, what had a big enough mouth to cause such grooves. On the dry plains where *Triceratops horridus* lived there also lived its hunter, *Tyrannosaurus rex.*

• • • • •

Julian breakfasted on smoked *Triceratops.* It was excellent, and not as leathery as he'd feared. After a long drink from the river he wrapped up the rest of the meat in the skin, slung it over his shoulder, and continued west, upriver.

He had walked for about an hour and was considering stopping to rest when something made him pause. It was a muddy footprint, clearly outlined on one of the wide, flat stones along the riverbank. As a paleontologist, Julian had good training in the study of footprints. But no expertise was needed here. It was the outline of a human foot. He knew immediately from the size that it was Dr. Shanker's, not Yariko's.

The feeling of relief was so intense that he sat down on the ground, right where he was. A human footprint meant that he was not the last man on earth. He was not alone; they were searching for him.

A second look at the print brought up the obvious question. Why was Dr. Shanker walking barefoot?

The mud was dry, turning to dust as Julian touched it. They were clearly hours ahead of him. They: but was Yariko there too?

Horrid possibilities came into his mind. Maybe Yariko had drowned trying to save him, and Dr. Shanker finally gave up and walked away. Maybe she had been killed by the *Triceratops,* maybe . . . Julian's heart was almost racing and he felt an irrational fury at Dr. Shanker for being alone, for being barefoot, for not taking care of Yariko.

But Dr. Shanker might not be alone. Surely Hilda was still with him, yet there was no dog print. Julian began to calm down again, and even feel excited. He was on the right trail. Now he only had to catch up with them.

The footprint pointed away from the river, toward the bushes. Looking closer, he saw that a path had been beaten through the bracken, and he followed it to the more open ground beyond. The trail continued through the low scrubby bushes and ferns, past a clump of trees, finally reaching a pile of rocks in the distance. Shanker must have climbed to the top of the rocks to get a better view of the land. Julian strode along the path, his excitement mounting in the hope that his friends might still be there.

The sun was high and he was soon drenched in sweat. There was hardly any shade in the field of scrub. The light glared down on his already aching head; bits of dirt and leaves stuck to his legs, and the heavy smell of crushed fern and herbs made him dizzy. After twenty minutes Julian was desperate for a rest.

He walked toward a stand of old bent-over juniper. A few logs lay on the ground in the shade. It was too tempting to pass up, and his friends may well have stopped there too.

He took a step closer, and one of the logs moved. The animal, whatever it was, seemed enormous, camouflaged in the dappled light. Julian gaped at it in disbelief. He couldn't see the shape well, with the intervening trees and the distance; but it was not *Tricer-utops*. The open terrain seemed ideal for a large predator such as *T. rex*. At that thought Julian turned and hurried away, heart pounding, using the spear to beat through the thick tangles of bushes. In his battered state he could not have run from anything. Even walking took more effort than he thought possible. Now and then he glanced over his shoulder, but nothing came out from the trees. If *T. rex* it was, it either did not see him or did not care to dine on him just then.

Finally he reached the pile of rocks, some forty feet tall. He could not imagine what had made such a stack. In the Quaternary, it might have been a glacial deposit. In the Late Cretaceous, nothing came to mind. It was composed of boulders tumbled together, each one as high as his knee. The ascent was steep, but shrubs and creepers had grown out between the cracks and made convenient handholds.

Julian began to climb. Halfway up he stopped to rest, sitting on a projecting rock, his shoulders drooping and hair sticking to the sweat on his forehead. His legs ached and throbbed. The clump of juniper was still visible; but he felt safer now, the sides of the hill being too steep for a large dinosaur to climb.

When he'd caught his breath and wiped some of the stinging sweat out of his eyes, Julian picked up the sack of dried meat and the spear and continued the grueling climb. The last ten feet nearly stopped him; it was a vertical wall of boulders. Several were loose, and when he clutched at them they turned with a scraping sound.

He hesitated before going on. Then he thought of Yariko and Dr. Shanker climbing up to have a look around, maybe even camping for the night. He stabbed the four corners of his hide bundle so it was securely fastened around the meat, dangling on the end of the spear, and reached up to the wall of rocks.

The first boulder he put his weight on instantly came free, throwing him off as it crashed down the slope, knocking several more rocks loose on its way. Julian lay sprawled where he'd fallen, clinging and shaking, while the splintering crashes went on, forever it seemed. At last he looked up. The cavity left by the boulder looked ominous.

When he'd calmed his heart again and regathered his courage, Julian slung the sack and spear over the top and crawled gingerly upward, testing each stone before putting his weight on it. As his head rose up over the top he saw a flat area, a rim, stretching away on either side, but only a few feet thick. Finally, when he was in the most awkward position, raised up to the level of his waist with his stomach pressed painfully against the edge, he saw what lay beyond the wall.

His mouth dropped open in shock.

A small field lay ten feet beneath him. It was perhaps twenty-five yards across, encircled entirely by a wall made of haphazardly stacked rocks and the occasional log jammed into a gap. The floor was smooth, covered with dung and dust, bits of stone cropping up here and there, a few bushes straggling in the thin layer of soil.

Near the center a dead tree stuck up, nearly stripped of its branches, only a twisted trunk with one side branch. Lying in a cluster around this, as if, absurdly, trying to get into its shade, was a group of six animals. They looked like cattle. They seemed tiny—Lilliputian—meek, light brown, perhaps the size of cattle. They were four-footed creatures with no horn, no frill, no weapon, but by the shape of the body and the hooked beak on the tip of the snout Julian could tell they were ceratopsians. Exactly which puny relative they were of the great *T. horridus* he did not know. The three that were facing toward him looked up with lazy expressions and did nothing. One of them flicked its head, as if it were chasing away an insect.

On the far side of the enclosure, against the wall, was a shelter made from great slabs of stone. The hut was thatched on top with sticks and dried ferns. It was man-sized and clearly too small to be a stable. It could not have fit even one of the ceratopsians, miniature though they were.

Julian's capacity for surprise should have long since been blasted away; but he was shocked by this scene. He had accepted the reality of the strange Cretaceous world, and what he now saw was impossible. Never mind that he himself was an anachronism: Julian's paleontological sense told him that bipedal, tool-using primates would not evolve for another sixty-five million years.

Students often ask me what animal is the most fearsome of killers. One serious contender is Deinosuchus, *the fifty-foot Cretaceous crocodile.* T. rex, *though not as long, was probably faster and more terrible. But to me the answer is obvious:* H. sapiens *must surely be the most fearsome animal yet.*
-*Julian Whitney*, Lectures on Cretaceous Ecology

– FIFTEEN –

1 September
8:29 PM Local Time

When Earles strode into the lab, the two physicists were once again side by side bowed over the computer keyboard. At her entrance they started and turned their heads so perfectly together that Earles had to smile. They looked like children caught in some naughty act.

"All right," she said, taking a seat beside them. "I know very little of modern physics, but I understand concepts well—and I have a good imagination."

The two scientists looked at her without speaking. They didn't know quite what to make of this chief of police, and hadn't from the beginning. Bowman had pictured someone like Hann: big, masculine, unimaginative. He couldn't decide if Earles should be taken seriously or not.

She certainly took herself seriously. "So: these people were creating matter—twigs, beetles—inside a sealed vault that not a particle of dust could have entered. I know: I checked it out."

"Matter can't be 'created.'" Bowman said. "That's not what they were doing."

"Exactly." Earles went on, undaunted. "Those things already existed. They came from outside, obviously, but they were picked up

178

and moved; I have no idea how, but then you're the physicists, not me." She looked at Ridzgy.

"We're trying to figure out how." Ridzgy sounded tired. "We've started getting the wiring back together in the vault. The extreme temperature was localized to one small part of the circuit, fortunately."

"There's more," Earles interrupted. "If stones and insects can be 'vanished' from one place to reappear in this vault, then couldn't the opposite happen? Couldn't something inside the vault vanish because it had been instantaneously transported somewhere else?"

Bowman got up and began pacing the room. "I'm a physicist, not a science fiction writer. You asked us to figure all this out for you, but you won't leave us alone to do it."

Earles ignored him. "If a beetle appeared inside the sealed vault, the air pressure would change. It would increase, by a tiny amount. Was that change measured?" She looked from Bowman to Ridzgy, but neither answered. "You said the twigs and such were all weighed after they appeared."

"Yes." Bowman sounded cautious.

"Let's say a pebble weighs one gram. It appears in the sealed vault, and causes a change in air pressure that the instruments can detect. From that, you could calculate how much mass had to disappear to cause the vacuum that pulled the door shut."

Ridzgy was looking at Earles with a new respect; but her first words were a correction. "It isn't mass that's important, it's volume. What matters is the volume of air something displaces."

"Volume, then," Earles said.

"Well, there are measurements for about twenty samples, including some small rocks. Size, from which we can calculate volume, and change in pressure in the vault . . . yes, there should of course be a predictable relationship."

"But what would that tell us?" Bowman sat down again, somewhat unnecessarily close to his colleague, Earles thought. "And the door wasn't sealed when the vacuum was created. Some air from the room must have rushed in before the door closed."

"We'd know if the volume could equal four-and-a-half people and a dog. Even if it was a rough number, we'd know if it was possible." Earles got up and walked over to the vault, where she looked in at the smashed dials and open, dangling wires. "Maybe they can be brought back from wherever they are. Maybe they can be retrieved, if they're still alive, just like they retrieved their beetles."

She turned back to the scientists. "I'll call the department's electrical engineer to help you with the wiring. Better yet, I'll get that student, Mark Reng. He said he knew all about the electronics of this thing. I want this place back up and running before morning."

• • • • •

Julian struggled to the top of the wall and sat, staring, legs dangling over the edge. After a moment the sun on his neck and the painful heat of the stones reminded him that he had to act. He threw the bundle and spear to the ground and scrambled down after them. Then he gripped the spear in his right hand and stared around at the corral.

Something about it seemed well-ordered, lived in, although it was hard to choose any detail that gave the impression. Perhaps it was the animals standing like cattle chewing their cud; or maybe it was the dry, well-beaten earth of the ground, and the neat stacks of wood here and there.

Making a sudden decision, Julian set out across the corral to the building at the far side. He passed within a few yards of the ceratopsians. They gazed at him out of wide brown eyes. None of them got up. They did not seem concerned.

The doorway was merely a gap between two stones. Julian looked cautiously inside and saw that the interior extended backward for several tiny rooms. The first room contained an ash-filled pit and a mound of hay built up against one of the walls; ferns, that is, and shrubs that had been cut down at the base. Some part of Julian's mind was satisfied, since grass *per se* had not yet evolved.

On the far side of the room another doorway led into darkness. He stepped over and looked inside. The roof of this inner room must have been made of stone or more neatly thatched, because it let in no light and his eyes were half a minute adjusting. The light that filtered in from behind showed a large chamber with a soil floor. A few large rocks lay about, possibly meant to be seats. In the center, hanging by a rope strung through its jaw, was the carcass of a small ceratopsian. It was half skinned, and the leather hung down over its back legs and dragged stiffly on the floor. The blood was already partially dried, and the carcass had begun to stink in the heat.

Stepping around it, Julian found another doorway on the opposite side. The third room was better lighted. It had a window, again only a gap between the stones, through which a shaft of light stabbed onto the floor. Beside the window stood a wooden barricade made of thick logs. Julian guessed that an opening led to the opposite slope of the hill.

In the corner, so wrapped up in shadow that it might have been missed, was a wooden couch, a length of tree trunk that had been split to form a smooth, flat surface. Stretched out on the couch, his back propped up against the stone wall, was a man. He was breathing quietly, asleep it seemed, his eyes only pits of shadow. He was wearing a tunic of cured leather and tight pants down to his knees, and his feet were bare. His hair was yellowish gray, long and tangled, falling about a gaunt, brown face that looked like chiseled wood.

Julian stared at him for a long time while his brain told him, idiotically, "*Homo sapiens.*"

As Julian's eyes grew accustomed to the darkness, he realized that he was being quietly watched. The man was awake.

After a moment he spoke. "I saw you come up from the river," he said. The voice was deep, measured, and quiet. The accent was strange and it took Julian a moment to decipher the words. "I have watched you." He nodded toward a chunk of wood in the opposite corner of the little, skewed room, as if inviting his guest to sit down.

Julian realized that he was leaning heavily on his spear; his head was spinning, and in another moment he would have keeled over in a dead faint. He sank down onto the seat, still tightly clutching the spear, as if the wooden solidity of it held a part of his sanity.

For a few minutes neither moved or spoke. Julian couldn't make his voice work, and the strange man seemed content to observe in silence. Finally he rose and stood looking down at Julian, the dim shadow of his broad shoulders falling across the room. Then he strode through the doorway.

When Julian felt he could stand up without his legs trembling he followed the man back through the series of rooms and into the dusty corral. The ceratopsians stood placidly around the tree, chewing wisps of fern. Their master had grasped a crude wooden pitchfork and was tossing clumps of hay—dried ferns of various species—in their direction. He threw a few more forkfuls and then turned to face his visitor.

The man's face, in the sunlight, was the color of old leather, smooth except for the corners of the eyes, and his chin was sprinkled with a thin, sparse beard. Something about the face struck Julian with a sense of familiarity, of comfort almost. Surely, he knew that face. . . . He shook off the odd feeling. Of course the man looked "familiar"; he was another human being, and Julian hadn't seen more than two of his own species in over a month.

"The others are gone," the man said, softly and with finality.

"You saw them?" Julian cried, excitement surging through him, once he understood the words.

The face took on a puzzled expression, the creases bunching around the eyes.

"Who are you?" Julian said.

The man gave him a searching look, and then, apparently deciding his visitor was harmless, wiped a hand on his thigh and held it out. "Carl," he said; or so Julian heard it. The accent was odd and the name could have been different. It sounded more like "Corl."

Julian took his hand automatically. The palm was rough, crossed with scars and calluses, and the finger nails were filthy. His grip was surprisingly strong.

"Julian Whitney. But. . . ."

It occurred to Julian that he might be suffering a heat-induced hallucination. Maybe the strain of climbing the hill had finally done him in, after the river and hiking in the heat. How else could he account for this Maastrichtian anomaly, this man who strode up and shook visitors by the hand? He felt like laughing. Here was the welcoming committee of the Late Cretaceous. But where had he gotten his twentieth-century manners and learned his English?

Julian looked his host over closely and decided that he was too dirty and wiry to be a mirage. He hoped he would dream up a more angelic hallucination.

Carl was silent.

"How did you get here?" Julian finally stammered.

"Walked."

"Walked . . . impossible. Don't you understand? Humans don't exist here!" In his state of shock, the irony of this statement was completely lost on him.

The man gave him a level stare without expression. Julian stared back, but he could not hold that gaze for long and soon dropped his eyes. It was then that he noticed Carl's left hand. It had only three fingers, the third cut off at the top knuckle. Twisted pink scars had grown over a good portion of the palm and filled in the space where the last two fingers should have been.

Carl walked calmly back to the hut and disappeared into the shadows. Nothing about the man had seemed threatening so far, but he was hardly friendly either. His expression was unreadable. Julian wondered if he were fetching a weapon.

Carl came out with a sort of shovel, roughly carved from a single piece of wood. He handed it to Julian without a word and then grasped his pitchfork again.

He began tossing hay toward his dinosaurian livestock. After a few heaves he glanced back at Julian and said, "We will talk after the work."

He seemed to be saying, stop gawking and start shoveling. Julian shrugged and fell to.

For the next hour they shoveled ceratopsian shit. The scattered clumps had to be gathered into a mound and then shoveled into a depression on one side of the corral that drained to the outside. It was heavy work but Carl seemed tireless for all his age. Julian was soon out of breath.

At first he was shy of talking to this impossibility, this apparition; and Carl did not break the silence with conversation. Once he glanced at the sky and said, "I eat when the sun sits on the wall. After dark, I go out." He had a most economical way with words.

Julian took in the stringy physique, the sharp features and skin that was neither light nor dark. A man perhaps between sixty and sixty-five; not so old, in fact. And in better shape than I am, he thought, or at least more practiced at shoveling shit.

The sweat ran down his chest and back, and his shoulders began to ache. The sun was a long way from the rim of stones. Finally he screwed up his courage and asked if there were any other people in the area.

Carl shook his head. "The others are gone, long ago." Then he added, "Your friend went west."

Julian's shovel stabbed into the ground. "You met them!"

But Carl shook his head again. He took his time, prying up a clump of dung that had dried into the crack between two boulders. Then he said, "I saw one human print in the mud. There was another?"

Julian described his companions, but Carl seemed puzzled over the concept of dog. Finally he said, "It was the one you call Shanker. A heavy person with a large foot."

Julian gestured with the shovel toward the west wall of the enclosure. "We were heading for the hills. They must think I went on without them."

Carl stopped again, leaned on the pitchfork, and stared at Julian in a searching manner, as if a new thought had come into his head. His blue eyes were disconcerting. It was impossible to tell if he was friendly, hostile, or merely indifferent. "West," he said, as if it had some meaning; but he added only, "they have a hard journey."

That seemed to exhaust his vocal capacity on the subject. He nodded toward another depression in the ground and said, "That could be cleaned."

After staring for a moment, Julian went to investigate. It was a stone-lined water trough, slightly recessed under the wall. A thick layer of mud and ferns covered the bottom and smeared the sides. It smelled so vile that he hesitated over shoveling into the muck. He stood a moment, frowning at it, until frustration got the better of him. What right did this being have to put him to work? As far as Julian was concerned, Carl didn't even have the right to exist.

He turned and blurted out, "You tell me who you are, and if you know where my friends have gone."

Carl only said calmly, "The work must be done." It wasn't an order; it was merely a statement of fact. Julian began to see that his host did not mean to put him to work. Carl simply did not know what else to do with a stranger. He was as confused about Julian as Julian was about him.

The very strangeness of Carl had begun to frighten Julian and he seriously considered throwing down the shovel, scrambling over the wall, and disappearing into the scrubby landscape as quickly as possible. Maybe if he ran away, Carl and his animals and enclosure would all disappear, and the world would make sense again.

But he looked at Carl and watched the broad, tanned back

bending and unbending as he worked. His movements were so calm and deliberate that, quite suddenly, Julian trusted him. The man might be the devil who had come back in time to haunt a lonely human; but he was a charismatic devil, who seemed to know exactly what he was doing. As rational thought began to return, Julian felt first a great relief at not being the only human being in existence; and second, creative, if unlikely, reasons for Carl's existence suggested themselves to his mind.

As he pondered, he returned to the problem of working the shovel into the recessed trough; the wooden blade was a few inches too wide. When he glanced back at Carl he found that he was being watched.

"Use your hands," Carl said, curtly.

Julian looked down at the mess in the trough and started to laugh. He suddenly felt better about everything. Mucking out cattle stalls had been a childhood job one summer long ago. Now, this simple necessity made him feel as though order had come back to the world. He leaned the shovel against the wall, got down on his knees, wrinkled his nose for all he was worth, and plunged his fingers into the mud.

As he scooped up the last of the muck, Carl's shadow fell over him. The sun was a handsbreadth above the rim of stones. He stood, wiping black mud mixed with ceratopsian shit on his jeans.

Carl nodded once, which Julian took to mean, "Clean enough." Then he reached into the trough and pulled a leather plug from the wall. Water gushed in, surprisingly clear. When it had risen to the top of the trough, he replaced the plug. As Julian found out later, the wall was riddled with wooden and bone pipes, reservoirs, filled up with rainwater.

They washed their hands in the trough until the water was black, and Carl drained and refilled it with fresh water. Then Julian followed his host inside.

Carl said nothing but simply went about his work. He carefully removed the ashes from a mound of hot coals. Then he lay twigs on the embers and fanned them into a fire. Next he set up a wooden tripod, blackened from use, and hung a clay pot of water over the flames. Julian noticed that his left hand seemed to give him no trouble, and he used it as much as his right; the injury must have occurred many years ago.

On a flat rock beside the fire pit Carl set a chunk of raw meat and a pile of fiddleheads. Julian smiled when he saw them: fiddleheads are the tops of young fern shoots that, before unfolding into the fern shape, are coiled and resemble the end of a fiddle; certainly an appropriate Cretaceous vegetable.

Carl handed over a crude bone knife; Julian took it as an invitation to stay to dinner. While the wood burned down they cut the meat into strips. They steamed the fiddleheads, fried the meat with grayish-white roots in a flat clay pan, and finally mixed the whole together. Carl added a wedge of animal fat and a few pungent leaves pulled from a dried bunch hanging from the ceiling.

It was a fantastic meal, especially after the poor attempts at culinary art that Dr. Shanker, Yariko, and Julian had produced. At home he would have added salt, and used a fork rather than splinters of wood, but the ceratopsian fat and dried herbs could not have been improved on. The room itself was cozy: the fire glowed, rising and falling, intermittently lighting up the bunches of dried plants hanging about, a few clay pans hanging from sticks in the wall, and the tools they'd been using, now standing in one corner. It was a rough but comfortable home; and it gave Julian a sheltered feeling that he hadn't experienced in a very long time.

"But where did you come from?"

"West," Carl said. "From the hills."

They sat on rocks placed just outside the doorway, one to either side, relaxing after the meal. Carl took some time settling himself, chewing on a twig and glancing at the sky. He stretched out his long, wiry legs, picked at his teeth with his fingernail, belched with a look of profound concentration, and then sighed in relaxation. "From the caves," he said.

"The caves?"

He nodded but did not explain any further; instead he leaned his head back and closed his eyes. After a moment he spoke again. "And you, from the east?"

Julian wondered how he knew. "From the epicontinental sea—the Niobrara. We started on an island just off the shore."

"That is a new name for it," Carl said. "You have traveled long." He looked at Julian with a new interest, as if he were reconsidering how much of a hopeless lost fool his guest could possibly be. "I was told about it, but have never seen it."

186

"Who told you? Are there other people?"

Carl spat out a wad of soggy bark. "There were too few." He looked at the western sky where the sun was setting, then to the east where the moon would rise, nearly full in a clear sky. The temperature was already dropping and the cooler air was refreshing after the unsheltered heat of the day. "They told me of it, from the telling of those before. A few stayed and so we grew. But that was many years ago. I am the last." He did not seem at all sad; he merely stated it as a fact.

Julian watched him eagerly. "Who were they? How many?"

The question seemed to have little meaning to Carl. "They built the sentinel mound." He waved his hand to indicate the hut and the enclosure. "I built the walls. But that was a long time ago."

They were silent as Julian thought over the strange hints. Carl's calmness was baffling. How long had he been alone? Ten years? Or forty? Had another group of people traveled back in time? There was no other explanation. Furthermore, Carl clearly originated from relatively modern America; nineteenth or twentieth century, anyway. His speech, his tools, his little oasis of a farm in this vast scrubby landscape, all pointed to an origin similar to Julian's own. Add to that the coincidence of their being in the exact same location on the globe at the exact same time in prehistory, and it was clear that Carl had been introduced by the same happenstance as Julian: the vault in Yariko's particle physics lab.

Crazy as this conclusion was, any other explanation seemed wildly impossible.

But Carl seemed to accept the Cretaceous as his proper home. It sounded as though a previous generation had originally been brought here. In that event, the timing was not exact: Carl's ancestors had been placed into the Cretaceous of perhaps a century ago. Still, in the span of Earth's time, a hundred years was essentially nothing. A minor change of setting in the vault? Natural variability of result when dealing with huge time scales?

Julian shook his head. He could not solve the puzzle without help, and his host was not forthcoming.

Carl rose and gathered the bowls, put them in a leather pail, and walked across the enclosure to the water source. He'd had enough of talking, and had returned to his chores.

Julian stood in the doorway and watched the pink and orange

streaks spreading in the western sky. The air was warm and still. The sounds that came to him were quiet and unthreatening. Carl washed the crockery with a soft abrasive sound, scrubbing at the clay with a handful of tough plant fibers. A ceratopsian snorted, shaking its head; another rubbed against the dead tree. Nothing could be heard from outside the stone wall; only a vast, blanketing silence.

Suddenly Julian envied his host this peaceful home, this private castle in a world without noise, without city lights to take away the night and the stars—without six billion human beings.

At that moment the yard within the ring of stones felt more like home to Julian than his apartment back in South Dakota ever had. Certainly it was a haven in the carnivore-infested wilderness. But even if *T. rex* himself were to appear over the wall, he had the feeling, ludicrous though it was, that Carl would deal with him in his unhurried, practical way.

Suddenly a conversation with Yariko, one night in their tree loft, came back to him. We could populate a village, you and I. She had been right after all. Here were the remnants of her village of anomalous Cretaceous human beings. But who had started it? Misguided rescuers, not realizing they too could be trapped here, not getting the time exact enough? If so, did that mean that the lab and the vault were in working order, and might allow reversion?

When the vault came to mind Julian had to take a mental step back. His previous life was too remote, especially in his current exhausted state, to seem real any more; the ambition of reverting to the lab seemed an academic one rather than a personal one. For the moment, he wanted nothing more than to find Yariko and bring her here to share this peaceful life with him.

Carl stood before him with a quizzical look on his face, holding the pail with the crockery piled inside it. Julian realized he was blocking the doorway and stepped aside.

When Carl came out again he sat down on the rock and said, "The storm is coming up fast. Your friends will not go far tonight."

"I need to find them, as quickly as possible," Julian said, waking with a start from his idyllic thoughts as the urgency of their time window rushed in again. "They're headed west, and I'll have to catch up before. . . ." He stopped. Carl could never understand the details about their time window, now only two weeks away. "I need to reach the hills," he ended.

Carl nodded as if this information did not surprise him.

Looking at the man's impassive face, Julian had the absurd notion that Carl already understood. He wanted to tell this strange being everything and ask for advice. He wanted Carl to tell him what to do next.

But Carl only said, "The higher ground is hard to travel over. Your friends will not find a way there quickly."

"Do you ever go back to your caves?" Julian asked, wondering how far west they were.

Carl nodded.

"Then it's not impossible." Julian spoke to convince himself more than anything. He didn't want to think about the dangers, the small chance of finding each other in the vastness; the fact that his companions did not know, as he did from the fossil record, that they had entered the territory of *Tyrannosaurus rex.* "I'd better start tonight," he added. "The moonlight should be bright enough."

"Nothing will move in this storm," Carl said quietly. "You would do better to wait for morning."

Julian looked out at the dark sky, over the walls of the enclosure. No signs of a storm were apparent, but then he'd never been a good weather watcher. A few hours ago he'd been afraid of this man. Now he was hoping that Carl might point out the easiest way and give him some advice on the terrain. "How long does it take, to the caves?" he asked.

"Alone, eight days. With you. . . ." Carl spat again. "First we will collect feed to keep the animals."

Julian turned and looked at him. "You're—coming with me?"

But Carl said nothing.

Tyrannosaurus rex, *the monarch of carnivores, cavorting at the apex of the food pyramid, was probably the least abundant of dinosaurs. No ecosystem could have maintained a large number of them. Consequently, few have fossilized. Only twenty more-or-less complete skeletons have been discovered, mainly in the Hell Creek Formation in Montana. Everything we know about* T. rex *is based on this shocking paucity of data. One thing is certain: with a small statistical sample, we have not found anything near the biggest or oldest* T. rex. *With our small sample of twenty, we are unlikely to have found the rare statistical outliers, the huge ones, the true monarchs.*
-*Julian Whitney,* Lectures on Cretaceous Ecology

– SIXTEEN –

1 September
9:15 PM Local Time

The physics lab was eerily quiet as the two scientists ran their calculations. Small objects, from insects to three-inch rocks; yet the equipment was sensitive enough to detect the minute changes in air pressure within the sealed chamber. And minute they were, so that Bowman shook his head and declared they were wasting time on a layman's crazy hunch.

Ridzgy persisted. The idea was farfetched, but how could she dispute its possibility when she herself had first announced the mysterious "creation" of beetles?

The answer, when it came, was frightening. A volume equivalent to at least four adults had to have been removed to create the measured drop in air pressure. It was an instantaneous change that caused air from the outer room to rush in, slamming the door with it; once sealed, the vault had maintained its vacuum. But instantaneous though it was, the change had been recorded.

Bowman spoke first. "Could it be possible?"

Ridzgy looked at him thoughtfully. "It seems impossible. Four people and a dog, or rather four and a half people and a dog, to become non-matter, and perhaps be reassembled somewhere else?

190

True, that dead man didn't sound cut in half so much as . . . vanished in half. . . ." She shook her head and settled once more in front of the control computer.

"But the principle, Marla. The principle may have been shown by the small objects, the twigs and such, that appeared. So it should hold for objects of any size."

On the computer screen page after page of commands scrolled by. Marla Ridzgy watched closely. "A molecule leaves a molecular-sized vacuum. When the air rushes in to fill it the change in pressure in the vault is imperceptible. But if that much matter disappeared. . . . Why, it was probably 30 percent of the vault's volume. Can you imagine the effect of such a vacuum? Even if it is transient?"

"The vault wasn't sealed," Bowman reminded her. "Which is why the damage wasn't worse. And then there was the dead guard. The blood is sprayed all over the outside of the door, but there's none inside the vault, except those few smears from the other people. That guard wasn't cut in half by the door, and his upper half never bled inside the vault."

"Then," Ridzgy said, pausing the computer's scrolling, "if we buy this as a working hypothesis—just for the moment—this vault, because of the exact settings, became a translocation machine. It brought beetles in from somewhere else, alive, and it sent a group of people to that somewhere. . . ." She left the last word hanging. "Do you realize how big this is, if they've really shown it? If they publish this, and convince the world of physics that it's been done? Only," she added after a short pause, "they're not here to tell anyone what they've done. Right now, we're the only ones who know."

Bowman watched his colleague's face carefully. "If they were sent to where their beetles came from, perhaps that cop is right. Perhaps they could be brought back again, if they're alive, if the vault was working and the settings were exact."

Ridzgy's eyes were alight. "Do we want them back?" she asked.

• • • • •

After sundown, in the moonlight, Julian climbed out of the enclosure, following Carl in search of ferns. It was not necessary to clamber up the wall the dangerous way he had done before; instead, Carl propped a wooden ladder against it, near the water trough.

191

Looking up, Julian could see the tips of great hollow bones jutting from the top of the wall. They were femurs of some huge beast, serving as water pipes.

As they climbed to the top he rapped his knuckles against one of the shafts, and a watery sound echoed up. How had Carl managed to drag such monstrous bones there and mount them upright? Julian had a ludicrous mental image of Carl striding over the rocky hills, an eight-foot, three-hundred-pound femur balanced on his shoulder.

They scrambled down the slope. Carl moved silently, but Julian slipped and stumbled on the half-seen stones.

At the base of the slope they headed north toward wooded ground, following a well-worn trail through the brush. Carl moved with a long swinging stride and was soon several yards ahead. The moon was full and lit up the open landscape.

The afternoon spent in safety had confused Julian and dulled his sense of watchfulness. He could almost have been back in Pennsylvania, fossil hunting with his father, if it weren't for the renewed sense of urgency to go west after his companions. He tried to calculate how far they might have gone since their separation. But that only made his chest feel tight, so that he had to think about breathing. Much as he wanted to think they were still searching for him, he decided they would have gone on, impelled by the short time window, knowing it was their only chance.

The bracken rustled nearby and Julian paused. Then, from the ground at his feet, a horrible sound broke out: a piercing, rapid, "Kah kah kah kah kah!" that seemed to echo all around. He stumbled backward, clamping his mouth shut over a yelp of fear. Twigs snapped, something scurried through the undergrowth, and there was silence again.

"It is better to stay watchful," Carl said, calmly looking back. "Some animals will attack without such warning."

"What was it?" Julian asked, curiosity getting the better of his initial terror. He stirred the bracken with his foot, but could see nothing in the moonlight. The branches and leaves threw confusing shadows.

Carl used his hands to indicate a small animal, about a foot long. "We might have killed it for tomorrow's dinner. But let it go. We have too little time before the storm."

Julian still saw no signs of a storm, but followed his guide closely, chastened and more attentive to his surroundings.

They skirted the wooded area and stopped in a boggy depression where ferns grew thick and immensely tall. Carl pulled a short curved blade from his belt—it looked like a huge tooth, filed flat—and pushed in among the ferns. Julian could tell where he was only by the tossing of the fronds as he hacked at their stems. A moment later he reemerged with a thick bundle, which he tied with a leather thong and lay on a dry patch of ground. Julian pulled out his own small knife to help.

By then even he could sense a change in the weather. The wind moved high in the branches and hurrying clouds obscured the stars. The moon took on a sickly glow. Suddenly Carl's three-fingered hand touched Julian's shoulder, and with a nod of his head he indicated the ground in front of them.

The mud was scored by three long depressions, close together at one end and fanning outward at the other. It was the print of a clawed, tridactyl foot. Although Julian had a good eye for tracks, this one took him a moment to understand. It was enormous, and his mental catalogue of animal tracks had to be rescaled entirely before his brain could even see the marks as a footprint.

"She is nearby," Carl whispered. He walked ahead into the open, his feet making hardly any sound.

Julian was a good woodsman; still he made a slight scuffing over the uneven ground. A stick cracked under his foot, and at the sound a dark object reared up over the ferns, about forty feet away. It was an animal's head. He saw it for an instant in profile, and it was longer than he was tall. Then it turned toward him and he looked into its wide, mottled face and its eyes; or rather its right eye, since his gaze could not take in both at once in that vast head. The eye glittered in the moonlight, peering down from a height of twenty-five feet. The huge jaws were slightly open, showing teeth as long as Julian's hand. Those jaws could have engulfed him whole, leaving only his feet sticking out.

Julian knew that he was, quite literally, looking into the face of death; a curiously expressionless face, the color of bark, dented and ridged.

It could only be *Tyrannosaurus rex*. He knew this partly because of the distinctive wide cheekbones and the stubby, two-clawed forearm, no bigger than a human's. But mainly, he knew because of the impossible size of the thing. Nothing had prepared him for the living animal itself: not the skulls he had studied, not museum skeletons, not movies or imagination clothing those museum models with flesh.

He had never before realized how well the name fit the animal. Several times in the past month Julian thought he might have glimpsed a *T. rex*, but now he knew that all the previous sightings had been of lesser animals. There was no mistaking the monarch of all dinosaurs. He stared up at it in disbelief. Then it lunged toward him. The speed of its leap was terrifying, and all paleontological and rational thought left him as he turned and ran.

He sprinted a few yards and then became mired in a deep mud hole. It would have been better to lie still, but he lost his head and began to thrash and struggle hysterically. He glanced over his shoulder, wondering why he was still alive, and saw a small dark object whiz through the air and smack the animal on the snout. The *T. rex* stopped, with a look like a dog hit by a rolled-up newspaper. Another rock hit the mud behind it, and then another beyond that.

There was an agonizing moment when they seemed frozen in place, neither man nor dinosaur moving, like actors on a stage waiting for the curtain to fall. Julian thought his heart would explode. In reality it must have been a second in time; but it felt like the longest hour of his life.

The great head swiveled away. It showed a bumpy profile and then the vast bulk of the body as it turned. From nose to tip of tail it must have been fifty feet long, and had Julian stood on Carl's shoulders, still he could not have reached its hip. As the animal turned he noticed that the left forelimb was mutilated: the claws had been cut off at the second knuckle joint, leaving the limb shorter than the other.

The animal was crouching nearly horizontal, its tail straight and stiff behind. It seemed to be sniffing with its nose to the ground; Julian could no longer see the head over the ferns. He watched the hindquarters move with surprising grace and a rippling of muscles under the brown hide. Then it disappeared into the darkness under the trees. There was a thunk of one more rock, a noise of trampled bracken, then silence.

Halfway up his shins in warm mud, Julian tried to twist his body around to look over his shoulder. The bundle of ferns lay behind him where he had dropped it.

Carl appeared from the shadows nearby, his bundle still balanced on his shoulder. He stooped to pick up Julian's ferns. "Easy to confuse," he said, tapping his forehead to indicate the animal's lack of intellect. Then he reached out with his free hand, clasped Julian's arm, and pulled him to dry ground.

Trying hard not to look as if his heart were about to burst, Julian forced a painful grin. "Faster than I expected," he said, the quaver in his voice unmistakable.

Carl gave him a quizzical look. "They are the fastest of all animals. You cannot out run the Big Ones."

"Will it—come back for us?"

Carl handed over the bundle of ferns and shrugged. "Perhaps. I am surprised she rushed you. She does not see well these days."

"She?"

Carl nodded. "When I first came here she was the leader. They live many years, longer than our kind. She is the biggest one I know, and the oldest. But the others drove her away when she became too old to hunt well."

If that charge with gaping five-foot jaws was not hunting well, Julian thought, I don't know what is.

He had studied the largest *T. rex* skeleton discovered to date, an individual estimated at twelve tons when alive. According to one hypothesis, fifteen-tonners were not uncommon. This matriarch was more than fifteen tons, if he was any judge at all. With the appropriate instruments he could have calculated her weight by the depth of the footprints and the viscosity of the mud; and as his mind turned over these possibilities, Julian realized that the fear was gone. Only the awe remained.

"You seem to know her," he said as they walked.

"We know each other." Carl shifted his bundle to the other shoulder. "She is more of a scavenger now. I set out carcasses for her at times, after I take some of the meat."

"You don't mean that you feed her? That might be what keeps her in the area!" The fear came surging back.

"I hope so," Carl said. He was silent for a few minutes as they trudged under their bundles, single file now along the narrow path.

"She is good to have around," he continued at last. "The other hunters stay away. They are smaller, but they are more dangerous to animals of our own size. No, I would rather face one of her kind." After a moment he added, "I feel a bond with her. Both of us are old, both loners. Perhaps she, too, has outlived everyone she ever cared for."

He said no more until they were nearing the hill of rock, the "sentinel hill" as he had called it. Then he turned to Julian and said, "I call her Corla."

By then the sky was almost completely overcast and the moon alternately appeared and vanished behind fast-moving clouds. The wind moved in gusts and carried an occasional drop of rain. Every few minutes Julian glanced about nervously, half expecting Corla's enormous head to emerge over the tree tops. But Carl seemed unconcerned.

The rain came down hard with a sudden blast of wind as they climbed the hill. Within seconds they were drenched, and Julian slipped and barked his shins as he carried the ferns up the rocky side of the hill. Carl moved about the top of the wall uncovering his wells, and Julian went inside to stack the ferns in the middle room of the hut. A few minutes later Carl came in, his bedraggled hair flat against his head and clinging to his shoulders. He added a few pieces of dry wood to the fire and the flames sprang up, sending shadows and lights spinning about the stone walls.

"Won't you douse the fire?" Julian asked, confused that Carl seemed to be settling in rather than preparing to head back out, and west. "I have some meat that I smoked," he added, lifting his bundle of *Triceratops* hide from the corner. "I suppose we won't have to carry water for a while yet." Now that the time had come, he was more than ready. Yes, Carl's home was comfortable after so long living in the wild, sleeping in trees and living from meal to meal; but time was moving, his companions were still lost, and it was past time to find them.

Carl's look was neutral rather than puzzled as he squatted beside the fire, carefully arranging more wood. "You may go," he said. "But you will be quickly lost. Weather in this season is a thing to fear." He raised his head and looked Julian in the eye. "I will find your friends, if they live. I will start when the sun returns."

Julian was momentarily at a loss. He realized he was counting on

Carl's help; and the thought of venturing out in the stormy darkness, alone again, solitary, leaving behind the only other human being—not to mention the probability of being stalked by a hungry *T. rex* accustomed to taking handouts—these thoughts shook his resolve more than he cared to admit.

"But. . . ," he shifted his feet but didn't release his bundle. "But daylight is hours away. I want to go now."

"Half the night has gone," Carl replied calmly. "Daylight will be soon. Sleep now. I do not wish to search for you tomorrow, and find you drowned in the flood that will come."

The wind gave a particularly loud howl just then and the sound of the rain changed. Julian shivered. Guilty as he felt being warm and sheltered while his companions endured this night in the open, he suddenly felt extremely weary. His eyes drooped and the bundle of smoked meat fell to the floor. He was half aware of his strange host quietly moving about, stowing things away. Then the scene became unclear, and he thought that Yariko was bending over him, pulling at him. "He's my father," she was saying. No, Julian tried to tell her, his hair is blonde. But he couldn't seem to form the words; and then Yariko was standing, and a cave grew in the darkness behind her.

Julian sat up straight, his eyes wide open in the blackness. The fire was a mound of glowing coals in a ring of ashes. He was sitting on a thick piece of hide with another on top, its musky smell filling his nose.

A dim shape stood in the doorway. The shape moved, and Carl's voice said, "We will leave early. If your friends stay clear of the river they will be all right. The Big Ones will never hunt in this." Then he turned and disappeared into the back room.

Julian lay for a time listening to the storm. The wind roared over the thatched roof and the rain splashed and pounded on the stones in the yard. A faint sound drifted in now and then, and his last half-awake thought was, "But how can there be music?"

The Rocky Mountains first emerged as a volcanic chain, the result of a subduction zone beneath the North American continent. The Pacific plate edge was in the process of colliding with, and sinking beneath, the expanding North American plate, just as it continues to do in our time. In the Cretaceous Period the Rockies had not yet been uplifted to their current grand height; for the Rockies are young, mere children as mountains go, unlike those ancient slopes in the east, the Adirondacks and the Alleghenies.
—Julian Whitney, Lectures on Cretaceous Ecology

– SEVENTEEN –

1 September
10:46 PM Local Time

The vault was clean and in order again. Broken dials were replaced and new wires, just brought in, were being fed through conduits in the ceiling. Bowman grumbled as Mark Reng worked inside the vault.

"There's no reason to hurry," he said to Ridzgy. "I'm tired, you're tired, that silent graduate student with his wires is getting on my nerves. I say we turn in, start again tomorrow. It's not like we're working against a time limit."

"We are, though." Ridzgy's eyes were drooping but she wasn't ready to stop for the night. "I want to know if our guess is true— before the government descends on this lab and takes it out of our hands. I want to make this translocation happen. Tonight is our chance. And as for that kid with his wires. . . ." She nodded toward the vault, where Mark's feet could be seen on the third step of a ladder, and lowered her voice. "There's only so much he can know, right? He'll think we're working to bring his bosses back. If we're careful, he won't notice the subtle differences in the settings."

"Well, you won't get far tonight." Bowman nodded toward the vault. "Oh, he'll have the wiring set up again by morning, and we'll be able to rerun the program. But it won't tell us anything about what really happened."

"Why won't it tell us anything?" The voice was not Ridzgy's. Both scientists whirled around and saw Earles. She had entered silently, and stood listening to the end of their conversation. "Is it that platinum bar that you need? I can get you another. Tell me the dimensions."

Bowman and Ridzgy exchanged glances.

"That won't work. We need the exact same. . . ." Bowman began, but Ridzgy interrupted in a rather loud voice.

"I've got the measurements for you here." She handed over a sheet of paper, torn from a lab notebook. Giving Bowman a stern look, she added, "Make sure it's exact to these dimensions. The wrong size bar, even a little off, wouldn't replicate the exact settings. And unless the settings are identical, we can't find your missing people."

"Very well," Earles said, taking the paper. "I'll set Hann on it." As she turned to leave she saw Mark Reng, with his crazy mop of hair, peering out of the vault's portal. He had an astonished look on his face that made her smile. Probably knew nothing of this platinum thing, she thought. They must have kept their secrets well, Miyakara and Shanker, for her only graduate student to know so little.

She felt sorry for him.

• • • • •

When the storm hit, Yariko and Dr. Shanker took shelter in a small clump of conifers. "Shelter" was perhaps too strong a word, but at least there was a semblance of a structure around them. It had begun with the wind, late in the day; the rain followed in the dark hours, and more wind. The trail across the brushy plain led straight to a dark smudge that, in the intermittent moonlight, could only be a stand of evergreens. This they hurried toward; but the breezy shower, which they were glum enough about, became a howling torrent before they reached the trees.

Under the branches the wind seemed easier and the rain was only scattered drops instead of sheets. They crouched, huddled together and shivering, Hilda trembling every time the wind gave a particularly loud howl or knocked down a branch.

It didn't last long. As the rain gradually stopped, Yariko began to explore their tiny woods. Soon she was shouting for Dr. Shanker.

"I found something. He must have been here. He's still ahead of us." Close beside a thick trunk, half screened by brambles, was a small, neat stack of wood. The pieces were about as thick around as Yariko's arm, and a few feet long; they were stacked log-cabin style, and held in the center were smaller sticks and split wood for kindling.

Dr. Shanker looked closely at the ends of the sticks. "Seems a strange thing for Whitney to leave behind. How in the world did he cut them?"

"I don't know," Yariko said happily. Her spirits were high now the storm was easing. "He must have made an axe of sorts, with a stone or—or a tooth. He knows all about teeth."

"I wouldn't have thought Whitney'd be so resourceful," Dr. Shanker said thoughtfully.

Yariko gave him an exasperated look. "Of course he's resourceful. He got this far on his own. But. . . ." She still couldn't understand why he'd gone ahead without them. Surely they would have found each other near the river, if he'd stayed there. Or did he think they'd gone on without him? That thought was enough to bring tears to her eyes.

"Nice of him, anyway," Dr. Shanker said matter-of-factly. "It's almost dry against the tree here. How'd he know which was the weather side?" He stood and felt his pockets for his keys. "Think these'll do as steel?"

They'd long ago each collected a small scrap of flint; now Dr. Shanker, by what miracle Yariko didn't know, managed to strike sparks and get the damp wood lit after several intense minutes. A partial imprint of his lab key was burned into his forefinger, probably for life as he said, but they had a fire. In the lee of the huge pine the wind was tolerable and the now light rain hardly touched them. Their biggest problems were the smoke from the damp wood and the steam rising from Hilda's fur.

Yariko fell to dozing as she sat cross-legged on the damp brown needles. People seemed to be crowding around her. But I had to lead you on, Julian was saying. To the caves you see. That's why I left you. Then Julian turned into Frank who said, Why did you leave me all alone you should have come with me. The air became swirling dust that choked her.

"Don't sit there," Dr. Shanker's voice said, and Yariko woke with a start as he shook her. Her eyes stung, and she coughed. "I scattered the fire," Shanker said. "The wind got the ashes—they went everywhere. Let's move."

Yariko got to her feet, still confused. It was very dark; she could see Dr. Shanker only as a bear-like form stamping out the last of the coals. "What's wrong? Why didn't you sleep?" She shivered. The wind felt colder.

"The clouds are coming in again. It doesn't look good." Dr. Shanker pointed to the patch of eastern sky showing through the trees, where the moon was again being obscured by clouds: thick, rolling black clouds, moving quite fast. "These trees won't be safe when that hits us. Let's get into the open."

Yariko was reluctant to leave the little woods; but Dr. Shanker said he preferred death by drowning to death by squashing under a fallen tree. They agreed to look for a thicket or a boulder so they could at least crouch against a partial wall. Dr. Shanker hurried out into the open, Yariko behind him.

But she wasn't the only reluctant one, it turned out: Hilda remained under the trees. She sat down and howled in fear, refusing to come out. They wasted valuable minutes trying to persuade her. Finally Dr. Shanker slung her over his shoulder and stalked off.

The rain fell harder. After some time Yariko realized she was following a sort of shallow depression, or trench. Perhaps ten feet to either side the ground rose gradually a good few feet, making her stumble if she strayed one way or the other.

"We must be in a dry streambed," she shouted, grabbing Dr. Shanker by the arm to get his attention.

"Probably joins the river," Shanker replied. "Whitney could easily have found it too. Convenient, to have an obvious path."

Yariko nodded, and concentrated on walking in the increasing wind. After a while she took a turn with Hilda draped over her shoulder. She did it more as a way to keep warm than as a favor to Dr. Shanker, for the dog was heavy, but it turned out to be a fortunate chance. They had spotted a dark patch, maybe bushes, in the last of the moonlight and were squelching through the rain, backs bent as the wind pushed at them, when Dr. Shanker suddenly cried out.

At first Yariko thought he'd stumbled and twisted his ankle; but the next instant he pitched forward and landed at full length in the mud. She couldn't see anything but his vague form.

"What is it?" she cried, struggling to lower Hilda. The dog clung to her in fear; thunder had been rumbling for some time, and now a louder clap than usual boomed over the drumming rain.

Dr. Shanker rolled onto his back and sat up, clutching his ankle and yelling. When the thunder grumbled itself out and Yariko could distinguish his words, what she heard was, "Attacked! It's got me!"

Yariko dropped to her knees and unceremoniously dumped Hilda in the mud. Then she crawled forward. Her heart was pounding and part of her wanted to run; but Dr. Shanker was still sitting up, and she couldn't make out another form in the darkness. She reached out a hand and touched his shoe.

"I've got it!" he said. "Goddamn lizard. Never knew something this small could hurt so much."

Yariko almost laughed in relief. "Where did it bite you? What was it?"

"It's here. Something the size of a chicken. Bit me on the ankle, and then I fell over it. Damn thing hung on forever. Think I'll keep it for our next meal—serves it right."

He sounded like his usual self. But when Yariko helped him to his feet he winced in pain and nearly fell over again. "Good thing you were carrying Hilda," he said. "That fall would've broken my neck with her weight."

"We'd better stop," Yariko said, loudly over the increasing wind and rain. It was nearly as bad as when they'd taken shelter in the trees, and it was rapidly getting worse. The moon and stars were now completely obscured; it was suddenly absolutely dark.

But Dr. Shanker wasn't ready to give up. "Not here. We're still in a depression—I can feel the water coming over my feet. I prefer life to death by drowning."

Yariko tried to act as a crutch but he said her pace was wrong, and her shoulder too high to lean on. "I'll use the spear," he said. "You go first. I'm right behind you. Just go slowly."

Yariko set out, walking as slowly as she could. Dr. Shanker's harsh breathing and dragging gait was loud behind her. Every few seconds she felt the air with her hand to be sure he was still there. Finally he told her to stop. "If you keep grabbing my crutch arm you'll make me drop the thing," he grumbled. "I'm not going anywhere you're not."

Then, with a suddenness that shocked her, the full storm hit. Yariko was knocked flat by the gust of wind that came over the plain, and the rain pounded her so hard as she lay on the ground that she wondered if she'd ever be able to get up again. The drops felt like solid pellets on her back and legs, or maybe like sharp splinters driving right through the skin. She realized that her face was in water and struggled to turn.

Something tugged at her legs and she scrabbled at the stony ground, digging her hands in as hard as she could. She realized she was lying on the edge of the depression, fortunately with her head on the high side. With an effort she managed to drag herself out of the deepening water, and the tugging on her legs stopped.

When she finally sat up she found it difficult to breathe in the seemingly solid sheets of rain. Cupping one hand over her nose and mouth, she felt around her for Dr. Shanker.

Her searching hand found fur; wet dog fur, and then an ear. She dragged herself up against Hilda and felt around in the blackness with the other hand.

"Dr. Shanker!" she yelled, but the sound came out as a strangled croak. She took a big breath through cupped hands and tried again, with better success. "Dr. Shanker!"

There was no answer. Was he unconscious? Was he lying with his face in the growing puddle, unable to move? Yariko grasped Hilda's tail in one hand and began to crawl in a slow circle, arm outstretched to sweep the greatest possible diameter. Her hand hit something hard and she grabbed it with a surge of relief; but it was only the branch of a shrub, torn from the soil and tumbling by. When she let go it was gone in an instant.

"Where are you?" she cried, miserably.

For a long time, hours it seemed, she felt around in the dark, not daring to let go of Hilda. Once she began to slip back down into the water and was nearly taken away by it, but she clung to Hilda and dragged herself out again. That was when she realized the dry bed had flooded, and the channeled torrent had swept Dr. Shanker away, just as the river had taken Julian.

Bits of shrubs and the occasional larger branch skittered by and even flew into the air around her. The storm was so loud that her senses were flattened, her mind confused. She lost all sense of place and of time, and at last collapsed on top of Hilda, clinging to the dog's solidity with all her might.

• • • • •

Julian woke to the sound of a steady rain. He had not slept well; his dreams had been tainted by the churning river, charging *T. rex*, and images of Yariko lost in the woods, calling his name as they had called for Hilda when she was lost.

A few embers glowed in the ash pit and a dim, gray light filled the room. The thatched roof had done a good job of keeping out the storm, but still there were a few puddles on the uneven floor, and the stone slabs of the walls were streaked here and there with moisture. He rose and pushed aside the dinosaur skin that covered the doorway. Outside, the corral was awash in thick brown water, churning in the heavy rain. The ground rose up highest at the center, forming a small stony island on which the animals stood, packed together around the dead tree. Bits of soggy vegetation hung from their mouths. There was no sign of Carl.

Julian shivered and turned back to the fire pit. There he found a bowl of clean water, and beside it, in a pot covered with a flat stone, something hot that looked very much like mashed potatoes. Two rough clay bowls had been set out beside the fire. He grinned; the lopsided bowls resembled the third-grade art work that he'd proudly brought home to his parents.

Huddling close to the warm coals, he used a handful of the water to scrub at his face and bristly chin. It would soon be time to at-tempt another painful shave with the pocketknife. He remembered Dr. Shanker laughing at him the first time he'd shaved— "I'm just waiting for you to hit that jugular," Shanker had said—and felt another surge of impatience to be up and searching.

Carl walked in, water streaming from his head and chest, his feet encased in mud. He shook the wet hair out of his eyes, blew his nose violently into his hand, and wiped the palm on his leather leggings.

Julian looked at him with concern. "Is the flooding bad?"

"The river has spread over the banks," Carl said. "My path is covered two miles to the east." At Julian's dismayed look he added, "We will still travel. The land is higher to the west." He stooped beside the fire and dished out the hot yellow mash.

"We should head out soon," Julian said. "How much time before the animals are ready?"

Carl smiled. "Not much time. Old men do not sleep as much as young men." He pointed to the ceiling of the middle room, where the carcass had hung the day before. "That was the last task."

"Will we take the meat with us?" Julian wondered how the man had handled the carcass by himself, but he didn't ask.

Carl shook his head. "It was a yearling. One of the adults kicked it and crushed its skull. The meat was already bad when I found it, but the skin was good."

"And the carcass?"

"For Corla. The Big Ones can eat anything."

Julian wondered what Corla would do without the handouts for a few weeks, and if she would hunt on her own. Old as she was, he could still imagine her bringing down a *Triceratops*.

After they ate Carl doused the coals with water, rather than covering them with ashes to keep them warm. The action seemed significant to Julian. Perhaps Carl planned to be away for some time. "Won't the enclosure flood?" he asked, picturing the result of a few days of heavy rain: a lake of muddy water, and the bloated bodies of the ceratopsians bobbing around in the center of it. But Carl said there were drains around the edge.

The rain was no more than a drizzle by the time they sloshed out into the brown pool. The water had already dropped considerably; the drains seemed to work. Carl had gathered supplies for the journey: sacks of smoked meat, two small skins (varanid lizards, Julian thought) for carrying water, and two light spears with fire-hardened points. They each took a food sack, a water skin, and a spear, and Carl carried a bundle of hides and an additional small sack that seemed to be elaborately stiffened with curved bones or sticks. He handled this one carefully, as if it was delicate.

They climbed the ladder to the top of the wall. It was well into morning already, but there was little light. Still, Julian was surprised at what he saw from the top of the wall. The river valley had vanished under a gray-white fog. The tip of Carl's hill rose above it, the only visible bit of terrain other than the tops of a few tall trees some distance away.

He turned for a last look at Carl's home. The pool of water had subsided and exposed the stony ground, choked with mud. He looked at the placid beasts, pulling mouthfuls from the bales of ferns that Carl had placed on the dry ground. He wondered if they would miss their caretaker. With a last look at the hut with its dinosaur-skin curtain, he turned and followed Carl down the hill.

They headed for the river. Water streamed along the ground and pooled in the cupped and curled leaves of the bushes. There seemed to be an unusual amount of animal activity; Julian heard rustlings and cracklings and the occasional eerie cry. He had the sense of an invisible world waking up. Holding tight to his spear, he followed Carl through the dense fog, keeping close so as not to lose him. Before long the river could be heard, loud in the dense air and sounding much closer than it turned out to be.

They came to it suddenly in the fog, and Julian hardly recognized it as the same river. It had turned into a muddy brown torrent, foaming and roaring, barely contained by its banks. Large branches, in some cases whole trees with tangles of roots still attached, swept past, crashing against unseen rocks. The rain had changed the landscape and must have washed away any traces that Yariko and Dr. Shanker had left behind. Julian leaned on his spear. He was disheartened and a little dazed by the thundering noise that seemed to come from all directions in the fog.

"We will find them," Carl said. He, too, leaned on his spear and looked at the river; but his expression was keen and attentive.

"Do you see something?" Julian asked.

But he said, "Only the river," and turned away.

They walked along the bank for nearly an hour, sometimes scrambling over stones, sometimes slogging through pools where the river had overspread its course. Then they plunged into a dense wood, and the fog thickened so that the great shaggy trunks of the pines loomed up, huge and solitary, then disappeared again behind them. Gnarled juniper trees looked like weird old men appearing suddenly and several times gave Julian a fright. If Carl noticed any of his jumps and starts, he did not show it.

Finally Carl slowed and then stopped near the river again.

"Here is where you crossed," he said.

Julian saw nothing. "How can you tell?"

"There is nowhere else to cross."

Fortunately, they were already on the correct side of the river; but it was no easy task following Hell Creek. They had to wade or scramble across each brook that emptied into it. The largest of the streams was a regular little river of its own, swollen with rainwater, and looked daunting to cross. Carl led Julian over a series of boulders, some of them partly submerged, the water foaming against them. Their placement was too regular, too convenient for chance.

Near the center of the stream both banks were invisible in the fog, and Julian could see only the patch of water directly beneath his feet, tumbling past. The motion of it threw him off balance suddenly, as if the ground were slipping away underneath him. He tottered, but Carl gripped his arm firmly and shouted over the roar of the water, "Better not to look down."

When they reached the opposite bank of the tributary they continued west, while Hell Creek veered away to the south. The sound of it sank into a vague hissing and rumbling. They were in an open area of stones and low scrubby bushes looming out of the fog. Julian began to shiver in the dank cold. Carl must have felt the cold also; he stopped and unrolled the bundle of skins. They turned out to be ponchos, one for each of them. Julian slid the skin over his head and found that it made a warm, if heavy, covering.

"The others can't be too far ahead," he said as Carl tied up the sack again.

"How long would they stay to look for you?"

Julian paused, thinking. He was still disturbed at the thought that they'd continued on without him, and wondered for about the hundredth time how they had missed him on the river bank. "I'm sure they would have searched that day, and maybe all night too," he said at last. "At most, I think they could be two nights and a day ahead of us."

Carl nodded; but Julian could not read his expression. His face was neither optimistic nor entirely grim. To Julian there seemed little point in searching for the others in the vast, soupy sea of fog. For the moment he forgot his own foolish desire to search at night in the midst of a tempest, and wondered what Carl thought they could possibly find before the fog lifted, if it ever did.

When Carl realized he was being watched, his expression softened. "They went this way yesterday morning," he said.

"They did? How do you know?" Julian tried to squelch the excitement he felt. Carl couldn't have seen any signs; but then he sounded so certain, as he did about everything.

"I tracked them before you found me." Carl pointed into the wall of fog. "The one you called Shanker. There may have been one other with him. He was making for a pine woods that I know."

Julian let the excitement take over. "Then they can't be too far ahead. Why didn't you find him yesterday?"

"I saw you. You were looking at my hill, and then you found my path. I turned back and waited for you."

Julian wasn't surprised that Carl had reached his hill first, but he did wonder how the man had seen him while tracking the others. Apparently, Yariko and Shanker had been very close by at some point, and none of them had known it.

They continued on their way.

Despite his new hope, the heavy silence of the fog began to weigh on Julian. After some time he thought there was a noise to the left, a scraping sound like rocks grinding together. He immediately thought of Yariko and Dr. Shanker walking; but he also thought of Corla, or some other terrible Cretaceous carnivore, stepping over the loose stony rubble.

Calling out loud was out of the question. He turned to ask Carl if they should creep up to the source of the sound and investigate; but Carl was gone. In those few seconds of hesitation Carl had continued ahead and been swallowed up by the fog. Julian hurried to catch up, but his foot snagged on a root and he fell hard on the rocky ground.

He lay still, holding his breath and listening, afraid that the sound of his fall would attract the creature that was moving in the fog. But he heard nothing. Finally he raised himself cautiously, wiped the wet grit from his palms, and looked around. He could not see beyond two yards.

"Carl?" he said, as loudly as he dared. There was no answer; only silence.

The dromaeosaurids, or raptors, were some of the smaller dinosaurs, but their brain size to body ratio was among the highest of all dinosaurs, indicating intelligence. These animals had a keen sense of smell and, like other predators, binocular vision allowed them to see in three dimensions, the better to hunt down their prey. It is thought that the dromaeosauridae hunted in packs, bringing down animals much larger than themselves. Velociraptor, *found in Mongolia, is perhaps the best known of the group; but there were equivalent species living in North America. Certainly their prey would not have known or cared about the small differences in morphology.*
—*Julian Whitney,* Lectures on Cretaceous Ecology

– EIGHTEEN –

1 September
11:05 PM Local Time

Bowman stared at his colleague. He had felt at ease working with her this long day; for that matter, he'd always liked her when they'd met at conferences. She was a no-nonsense scientist and also, he'd often felt, quite attractive in a mature, near-fifty kind of way. Even when he was married, he couldn't help noticing her; and now that he was single, he'd begun to have ideas. This happenstance that threw them together so intimately was just what he wanted.

But her behavior to Earles was inexplicable.

"Why are you staring at me? You know perfectly well what's going on," Ridzgy snapped.

Bowman watched the too-quiet kid, Mark or whatever his name was, slip out of the lab without looking at them. "I do? I only know you sent that woman off to get a useless piece of metal. You know perfectly well that a new piece won't work. Even with the correct dimensions, there'll be internal differences in mass and form. It can't replicate the original bar, and without that we can't replicate the settings."

"Of course we can't." Ridzgy was clearly impatient. "But we can still do what Miyakara and Shanker were doing—translocate objects and bring them into the vault. Does it matter if our collections don't come from the exact same place as their rocks and beetles?"

"But . . . but Marla, you told her we could try to bring Miyakara and Shanker back. You told the police. . . ."

Ridzgy made a dismissive gesture with her hand. "She doesn't know the difference. We'll get this thing set up and running by morning, do the experiment using the settings they recorded, and see what happens. If we don't bring the missing people back, who would blame us? It'd be a slim chance anyhow, even if that platinum standardizer hadn't been melted. Who's to say they're even alive?"

Bowman sat down heavily, staring at his partner. "But they'll come back, if they are alive," he said, expressing a new thought. "What's to stop them making it home eventually? Even if they're in the middle of the Sahara or something—surely they'll be found. Then they'll claim their work."

Ridzgy smiled. She had beautiful teeth; somehow the fluorescent lights brought them out. "They can try. But they'll have to be fast." She leaned closer to Bowman and lowered her voice. "What if we went somewhere? We could translocate ourselves, after collecting a few samples to see what kind of place we've locked on to. We take all the notebooks," she swept her arm across the lab bench, "and copies of all their files; and we destroy the original files, except the program that runs the vault. Then we get ourselves back home and start writing. And building; we can duplicate this setup."

"I admit it's a clever way of walking out with all the information," Bowman said. "But if we make a stir with our 'findings' it'll be obvious what we did. Everyone'll know."

"Who? The Creekbend South Dakota police force? A small-town woman who thinks she's Sherlock Holmes, and that idiot sidekick of hers with the cigarette breath?" Ridzgy laughed, and the sound wasn't pleasant. "Once we disappear too the university and OSHA will descend on this lab and dismantle it. They'll never let Miyakara and Shanker do this again, even if those two do eventually show up."

Bowman shook his head. "That graduate student who just left—he'll know."

"What ambitious young grad student wants to stay at the University of Creekbend, South Dakota? This place is a dump with no real funding. I'll give him the fellowship he's only dreamed about."

When Bowman didn't say anything, Ridzgy leaned closer again. "This is the chance of a lifetime. I'm offering you a full partnership." She leaned back and crossed her arms. "You aren't actually going to say no, are you?" Her voice took on the derisive tone that was particularly galling to Bowman.

"No," he said. "I mean yes. Yes, I'll do it."

"Then let's get to work." Ridzgy spun her chair around and opened a file. A large beetle, photographed in various poses, appeared on the screen. "The first thing to do is put in a disc and save this image for ourselves. Then, we delete it from the hard drive," she said, carrying out the actions as she described them. "We do the same to every file. Once we've taken all their notebooks with the descriptions and measurements, there'll be no clues where those people were translocated to. The University won't be sending out any rescue party to the Sahara."

•　•　•　•　•

The world was gone. There was only thick, gray-white fog pressing in on him in dank silence. For an instant, Julian thought his sight and hearing were gone too.

Then, without the least sound, a wall appeared out of the fog in front of him. It was not the shifting, amorphous gray of fog, but a solid form. It filled his whole range of vision for an instant, and then it was gone. Julian felt a faint stirring of the air around him and smelled a rich, mulchy odor of dung.

He began walking, slowly, trying to make as little sound as possible on the pebbly ground. He felt a kind of shuddering beneath his feet, barely perceptible at first. There was something, a blur, a sense of motion, caught out of the tail of his eye. But by the time he turned to look, there was nothing but empty fog.

Then began an endless nightmare of huge gray shapes appearing and fading away all around him, a trembling of the ground, blindness, and an overwhelming sense of insignificance—his own insignificance. A herd of immense creatures was on the move, and he was in the middle of its path.

Julian knew it must be *Triceratops*; few other animals moved in such large herds, or were so large themselves. He wondered if he'd be trampled or gored to death, or both; but either he was lucky or the creatures could see well enough in the fog to avoid him. They passed to either side, and he caught glimpses of their great, reddish frills and hooked beaks.

What direction they were traveling he had no idea; but he tried to orient himself with them and keep moving, so as not to stumble into anyone's direct path. It was the only thing he could do. Sometimes the herd swerved a little one way; other times they and their human flotsam drifted in the other direction. It was not long before Julian lost his bearings completely, and with it the last dregs of confidence. He trailed along, terrified, shivering, arms huddled together under his poncho.

After a while, hours it seemed, but it might have been only ten or twenty minutes, the creatures vanished as suddenly as they had appeared. He was alone again, lost in the fog.

He continued slowly up a gradual rise, to what must have been a ridge or mound, and stopped when he felt the ground level under foot. He wondered if any predators might be trailing the herd; if so, they'd find him an easy target.

At last the fog began to thin. A cold breeze found its way up into the poncho, but he thought it was welcome if it blew the fog away. He stood for a long time, shivering, and finally sat down on the ground. There was nothing to be done but wait for Carl to find him, or for the fog to lift so that he could find his own way.

As the fog thinned to a gray sky a stony, tumbled slope spread out at Julian's feet. The daylight was almost gone. The sun, pale and smeary, was squeezed into a thin strip of relatively clear sky between the clouds and the western horizon. Soon it would disappear behind the low hills.

Julian turned to look back the way he'd come. The fog still sat in the river valley, although here and there a dark smudge of trees rose up into view. He was surprised at how far and how high he had climbed.

"Now you will see something," said a voice behind him.

Julian wheeled around and saw Carl, leaning on a spear, gazing out over the plain toward the sunset. He nearly laughed in his relief.

Carl made no comment on his companion's disappearance; he only nodded toward the west.

Julian looked out over the rocky plain below him, misty but visible. The ground was a patchy greenish brown, the color of earth and bushes, but meandering through it was the wide, black, trampled path of the herd. Here and there great beasts had paused to rest. The bulk of the heard lay far ahead, visible only as a gray smudge. Not for the last time, Julian wished for a good pair of binoculars.

Then he caught a sudden motion among the stragglers. A young animal, only the size of a small rhinoceros, had wandered to the edge of a patch of trees and was now running back to the safety of the adults, its short legs pumping madly. As the uncertain breeze shifted Julian heard a high, piercing squeal. It sounded like a distress cry. Julian realized he'd just learned more in that instant than any paleontologist had ever known; for dinosaur vocalization is one thing that cannot be fossilized.

He was so engrossed watching the baby *Triceratops* that he nearly overlooked the tiny, insignificant thing that now emerged from the trees. It was laughably puny, shorter than a human and no bigger in body than Hilda. The head was birdlike with large, forward-facing eyes. The forelimbs were held well off the ground, and the long, curving claws were curled back toward the chest. The hind limbs propelled it forward with immense, graceful strides, the feet only rarely touching the ground. The run was much like that of an ostrich, and with each powerful thrust the animal seemed about to take to the air. It was a raptor; and had he been in Mongolia instead of North America, Julian would have said it was one of the *Velociraptor*. Of the North Amercian species, it was too big for *Saurornitholestes*, too small for *Dromaeosaurus*; but without a look at the skeleton, he would never know for sure.

He wondered how it could try to bring down something as large and well-equipped as *Triceratops*. Surely even a young one, if attacked head-on, would merely step on the little intruder and squash it. But the carnivore streaked past its prey and, wheeling, headed the young animal toward the trees again.

Julian now saw that three more were hunting with it. The attack was beautifully coordinated. Before any adult *Triceratops* could reach the scene, the animal was steered straight into the claws of the other three hunters. One leaped gracefully onto the youngster's back. Perched up there it looked ludicrously small; but where its claws were buried, streaks of bright crimson ran down the smooth hide. The youngster leapt and twisted, and a thin shrieking sound came up on the breeze.

Four of the adult *Triceratops* thundered to the scene; the attackers left their prey and vanished into the trees. But the damage had been done. The injured animal followed the adults for perhaps two hundred yards before it staggered, sank to the ground, and finally tumbled onto its side. The adults circled it a few times, grunting, and then lumbered off to join the herd.

"They'll come back for the meal," Julian said, thinking out loud. Then he added, "But it's a good thing they've already made their kill. I suppose we'll be passing right through their territory."

"And then into the territory of the next pack," Carl said.

That was not a comforting thought. Julian turned away and silently followed his guide along the ridge as the light faded.

He stayed close to Carl, not wanting to lose him again. But as they walked, his thoughts turned back to Yariko. She and Shanker had passed through a violent storm without shelter, and were now exposed to hungry predators. The image of the raptor skimming over the ground toward the young *Triceratops* replayed itself over and over in Julian's mind. Those small predators did not have the majesty of Corla; but because of the eerie, almost unearthly grace of their movements, they were more frightening, more the stuff of nightmares.

He was about to ask if they should stop and search for signs of the others when Carl suddenly stooped and looked intently at the ground. Julian knelt beside him.

In a small depression of mud and trampled leaves were the four-toed prints of a large mammal: a dog, in fact. They were smudged and blurred, probably by the fierce rain, but still clear enough; Hilda must have walked through a puddle, and the prints remained after the water dried up.

"Hilda!" he cried. "They were here. Let's hurry!"

Carl stood slowly, gazing down at the clover-shaped prints. "This is a strange creature," he said.

Impatient as he was, and wanting to dash off in hot pursuit, Julian could understand the man's confusion.

"It's a dog," he explained. "A large mammal—a carnivore. A pet." When Carl only stared at him, clearly not understanding the word "pet," Julian added, "Like Corla. An animal to feed who will then protect you."

Carl nodded as if he now understood. "This animal then would stay with your friends?"

"Yes. They'd be together." Julian looked out over the darkening plain, and then back the way they'd come. "It's amazing that we found these prints. In all this area, we walked right up to a low spot that still had some mud." He looked at Carl, wondering if it wasn't chance.

Carl smiled, or at least gave his version of a smile, very small and humorless although comforting. It reminded Julian of a particular look of Yariko's when she was preoccupied and not quite attentive to him, but still trying to be gentle and reassuring.

Carl turned away and began walking again with his long stride. "It was not chance," he said after a moment. "I tracked the man nearly to here and marked his heading against the trees." He raised his arm and pointed to a distant smudge of darkness, another clump of stunted trees. "I was looking for prints in low places," he added, as if that should have been perfectly obvious. "That is why we have been moving slowly."

Julian hadn't noticed the decrease in pace; obviously his idea of "slowly" was not shared by Carl. With renewed spirits, and renewed faith in his guide, he looked happily around at his surroundings.

A group of the *Triceratops* could be seen fairly close by, and in the last of the evening light Julian saw them engage in a strange behavior. One adult began it by flopping over on its side and then, unbelievably, rolling onto its back in the dust, legs waving in the air. A faint snorting sound came from it. Others soon joined in, grunting and sending up clouds of reddish dust. After a while they rolled back onto their bellies all at once, shaking their immense heads so that their long ears flapped audibly.

"Are they playing?" he asked, incredulous. "How can they do that?" Big animals just couldn't roll like that without injuring themselves; elephants were the perfect example. But he didn't mention them.

Carl gave him the usual expressionless look. "They are cleaning themselves," he said. "Removing parasites. Also, it feels good to them."

The animals were moving off now. Soon the ridge of land hid them from view, and Julian's thoughts, momentarily diverted by the feelings of *Triceratops*, went back to Yariko.

Hilda's prints had lifted his heart; but now he began to wonder if Yariko was with her. So far there'd been no signs of her, and Carl had only tracked one person, probably Shanker, the day before.

The clump of stunted trees was before them. Carl stopped suddenly and said, "Your friends stopped here."

Julian pushed forward into the trees and looked wildly around, as if expecting Yariko to pop out from behind a thick trunk. He saw nothing.

"Smell," Carl said, quietly.

The light breeze brought Julian the scent of mud, rotting leaves, a heavy animal smell wafting up from the trampled path of the herd; and, very faintly, ashes. There was no mistaking it, now that it was pointed out to him.

"Maybe they're nearby," he cried eagerly, hurrying forward.

Carl shook his head but Julian was already under the trees. He searched the ground for a mound of ashes, but saw nothing although the smell was stronger.

"They were here before the rain," Carl said. "The ashes were scattered in the storm. We too will stop here."

They sat down on the stony ground between the knobby trunks of bushes and took out strips of dried meat from Carl's sack. The meat was tasteless and leathery, but Julian was happy enough to put something in his stomach after such a long day. He had not eaten since breakfast, and neither, it seemed, had Carl. The water in the skins tasted terrible; he could hardly choke it down. After the meal, he propped his back against a tree and dozed off.

A soft scuffling noise startled him out of his sleep. Opening his eyes he looked down at a tiny lizard-like, or maybe bird-like creature, surely one of the smallest of dinosaurs. He sat as still as he could while it rooted around in the soil, perhaps looking for grubs, until at Carl's approach it darted away through the dead leaves.

"We must go now," Carl said. Julian stood quickly and shouldered his food and water sacks. Carl set off immediately, keeping within the shrubs and trees, walking fast.

They held the rapid pace until the clouds broke and the moon showed. By its position Julian estimated that the night was half done. Carl seemed especially watchful. He did not look about but rather seemed always to be listening. They continued even after the moon disappeared beneath the horizon. Julian thought Carl would stop to light a fire in the chill before dawn; but instead he paused only to drink from a cold stream and hand Julian a scrap of dried meat, and then, later, when the worn strap on his shoulder bag broke and had to be retied.

Julian was exhausted and hungrier than he'd ever felt by the time the stars began to fade. It was a clear, cool morning, and very still. The ground was sloping upward. Sometimes they passed low cliffs. He leaned heavily on his spear, trailing behind, silently begging Carl to stop and rest; but he said nothing out loud.

He realized suddenly that they were walking in an old stream bed. It ran more or less east-west, directly out of the western hills. It was strewn with boulders but made easier footing than the scrubby ground to either side; and it gave a path to follow. He would have liked to stop and study the exposed rock layers; but he was more impatient to catch up with Yariko. Carl would not have understood the delay, anyway.

"Do you think they went this way?" he asked finally, as he almost trotted to keep up.

"They followed this stream bed," Carl said, with his usual finality of expression. His next words were not so comforting. "But it flooded during the night. They could not have stayed in it long."

Julian stopped. He was breathing hard, and felt wobbly. "You mean we might not be tracking them anymore?"

In answer Carl held out his hand, palm up. Resting there was a wad of what looked like cloth. It was wet and dirty, and the color was hard to tell, but Julian guessed that it might be blue-jean material. It seemed to be knotted, and there were stains on it that could have been blood.

"Where did you find that?" he asked in amazement.

"Near some bushes," Carl said. "I have looked at every small shelter as we passed. There have been few signs of your friends."

Julian took the scrap of cloth, wondering why it was knotted. Yariko wore jeans; perhaps she had been injured, and had torn off a strip for a bandage. "Maybe this was washed away when the stream-bed flooded," he said, voicing one of his growing fears. "Maybe they were never even here."

Carl turned and set off again. "It was dropped after the storm," he said. "This night, when the moon was high."

"Then we're catching up," Julian said. He hurried after Carl.

In the first gray light before dawn they passed a low cliff, no taller than a man. It caught Julian's attention because it was not the granite he'd come to expect in this terrain. It was made of layers of sedimentary rock, and had clearly been uplifted and then tilted slightly, but in no way metamorphosed. He guessed that they were walking along an old fault scarp.

Peering at the exposed face in what light there was, Julian thought he could distinguish several layers of mud, some with impressions of shells, and one thin white layer. He scraped at this layer with a fingernail and tasted it. It was chalk: calcium carbonate.

Here in their billions were the calcium shells of tiny one-celled organisms, deposited over millions of years when the continental seaway had covered the area. On top of this thin layer of white lay a dark, compressed mud, perhaps indicating shallow water with salt marshes or sea-grass beds. Over the mud lay a thick deposit of what looked like volcanic ash.

There set out in successive layers was the history of the Niobrara Seaway. The slow transgression—the change from shallow to deeper water indicated by the chalk; and then in mirror image, the regression as the seaway shrank and left the region uncovered again. Volcanic ash could only have come from the west, from vast eruptions that sent their debris high into the atmosphere to rain down in a thick layer for many hundreds of miles around.

"The layers tell an interesting story," Carl said.

Julian turned and stared. He could not have been more surprised if Carl had begun to rattle off physics equations. "What story do you see in it?" he asked.

"What do you see?" Carl said. But before Julian could answer, he added, "We do not have much time. In a few hours we can rest."

They continued up the broad path of the stream bed. Julian felt certain that Yariko and Dr. Shanker would have kept to this path; it was the only obvious feature in the landscape, and certainly the easiest trail to follow westward. But there were no more signs. They found a good place to stop, a niche with an overhanging boulder and walls made of thick, tangled bushes. It looked ideal for a camp; but although he searched the ground, Julian saw no hint that anybody had stayed there. However, he was happy to put down his sacks and settle gratefully into sleep.

They traveled along the stream bed for two days, sleeping only during the darkest part of the night when the moon was down. The pace was grueling, although Carl let Julian stop to rest every few hours. Julian would sink down on the ground or a convenient rock, feeling as weak as if he'd been swimming through molasses.

Carl usually stood by patiently or walked a short distance ahead to spy out the path. By the third morning Julian was beginning to adjust. His muscles were not so stiff, and he had the strange sensation that he could walk forever. He realized he was beginning to turn into another leathery, taciturn Carl.

Early in the third morning the stream bed wound down into a gully. The bottom of the gully held a few stagnant pools of water and was nearly choked up with bushes and low-growing succulents. Spiders and small brown snakes seemed to have colonized this little muddy oasis; they were everywhere.

For two days there had been no signs of Yariko, Shanker, or Hilda. Carl said nothing; Julian questioned him constantly, impatiently, and was almost beginning to lose faith in his guide. It seemed foolish to walk on and on at such a pace when it was no longer clear that his companions were ahead. But where else could they be? Why would they turn off this clear path leading straight into the west? Reason told Julian that scraps of cloth or other items were remarkable things to find, and that footprints couldn't be seen in this dry, stony ground. There was simply nothing to track, and Carl must be going on assumption.

Up ahead, Carl suddenly ducked behind some thorny bushes, signaling for his companion to get down. Julian dropped to the ground and crawled over to join him.

"What is it?"

"Listen." It was barely even a whisper.

Julian heard something moving over the rocks. It made an uneven sound, not at all like a quadruped, but like a biped with a strange shuffling gait. There was a sound like labored breathing. Then something heavy fell to the ground, and, quite distinct in the still air, there came a string of curses.

Before Darwin's monumental publication, evolution was held to be a rigid sequence of steps preordained by God to produce humans. According to Darwin, there was no preordained endpoint to evolution, and therefore one species was no "higher" or closer to the goal than another. A bee is just as highly specialized and adapted to being a bee as a human is to its specific circumstance. Both are the current endpoints of their evolutionary lines, the current tips of the twigs on the evolutionary tree.
—Julian Whitney, Lectures on Cretaceous Ecology

– NINETEEN –

2 September
12:01 AM Local Time

Earles locked her desk, switched off her computer, and put on an old sweater over her uniform. Picking up the small box that Hann had just brought her, she went down the hall to the lobby. "I'm off," she said to the cop behind the counter. "Time to get a little rest. I'll be back by five. If any calls come in from those scientists forward 'em to my cell phone."

"Will do," the man said.

But Earles didn't get her few hours' rest, as it turned out. Before she reached it the front door opened, and Mark Reng walked into the station. He looked startled to see Earles. "Oh, I was just going to leave you a message," he said.

"No need." Earles led the way back to her office with a sigh. She unlocked the door and took off her sweater, placing the small box on her desk again. "Is the rewiring progressing well?"

"Yes, all that's fine. Maybe a few more hours and I'll be done." Mark held a small book in one hand and a piece of paper in the other. "I guess police chiefs keep hours like graduate students," he said with a nervous laugh.

Earles pushed a chair toward him. "Just say it right out," she said, leaning back against the desk and folding her arms. "This can't get much stranger than it already is. I'll listen to anything."

Looking reassured, Mark sat down. "About that platinum bar," he said, and stopped.

"Right here. Brand new, ready to go. I measured it myself to be sure they made it correctly." Earles opened the box on her desk and tilted it to show Mark the dull gray-white metal inside.

"The thing is," Mark said, "that bar won't work. I mean it'll work as far as getting the vault operating again, but it won't create the same results."

"Those physicists think it will. Are you saying they're wrong?"

Mark looked uncomfortable again. "I know this sounds terrible, but I don't think they want to find those people. I think they're intentionally setting things up to work but with slightly different settings. You see," and he looked up with almost a pleading expression, "they know as well as I do that a different chunk of platinum in the circuit will modify the results in small ways. They can't replicate the exact conditions without the original bar."

"Explain," Earles said. "And explain why they'd be so deceptive. What reason would they have?"

"Well, if my boss was really working on translocation of objects, that's a huge finding. She and Dr. Shanker will be famous when they write it all up. They'll be the first to show such a thing is possible. Science is a very competitive field, you know," he added, somewhat defensively. "I've been to enough conferences to know that physicists can be aggressive about their careers. If my boss never shows up, those two could steal all the results. That woman Ridzgy, has been copying files onto CD's, and that's not allowed."

"I had no idea scientists were so cutthroat," Earles said. "It sounds as bad as the criminal world. OK for motive, if you're right. Now tell me about this metal bar."

"The settings are predetermined in the program," Mark said. "But then there are the fine adjustments. Someone goes into the vault, just before the run is started, and watches the dials while making delicate adjustments. I've done it so I know what to look for; I could do it with a new piece of platinum."

"But you said a new piece wouldn't work."

"Not for those people you've brought in—they don't know the

details. They only know what's recorded, and the dial settings written down are for the original platinum standardizer. This new bar isn't the same as the original. The differences might be undetectable to us, but the fine adjustments needed are within those differences. They'd need the same bar, exactly as it was, to exactly replicate the experimental run."

"And they can't have the same bar," Earles said thoughtfully. "Yet she seemed confident, giving me these dimensions. She even said straight out that without the correct dimensions they wouldn't be able to find the missing people."

Earles didn't realize she was staring fixedly at Mark until he began to squirm. She was thinking of Bowman's astonished look when his colleague had handed over the measurements, and Ridzgy's interruption when Bowman was about to say something.

Earles focused on Mark again. "You think she deliberately lied to me?"

"Yes. I've been listening to them talking, from inside the vault. They ignore me. Kind of convenient, actually. And they've started erasing things from the hard drive of the master computer."

That impressed Earles more than anything else. "Erasing files? That's a criminal act. That could compromise the investigation. And the material doesn't belong to them."

"It will if my boss doesn't come back to claim it," Mark said, and his mild voice was surprisingly angry. "They'll run a perfect experiment when it's all set up, but they won't retrieve objects from the correct place—or time. Then they'll take off with all the records and their own new data, and publish it as their own."

This investigation was surely the strangest imaginable, Earles thought. If Mark was right. . . . She strode to the door, ready to charge into the physics lab and confront those people. But Mark spoke again.

"There's something else I think you should hear," he said.

• • • • •

Julian leaped up and hurried forward. He pushed aside a screen of bushes and found himself looking down at Dr. Shanker.

Shanker was on his hands and knees. He looked back over his shoulder when he heard Julian crashing through the twigs; by the expression on his face, he obviously thought he was done for.

222

When he saw Julian and then Carl behind him, the look changed from terror to amazement.

They helped him to his feet. He was so weak that they had to support him, one at each arm. He raised his head with an effort to gape at Carl, and then turned to Julian. But he said nothing.

In a more concealed spot near the cliff wall Julian eased him down gently onto the ground. He could not see any obvious injuries. Dr. Shanker sat with his head lolling back against the stone but his eyes were open and he seemed to be alert. "Well, Whitney," he whispered. "Did you come back for me?"

"What happened to you? Where's Yariko?" Julian felt like shouting and shaking him. If Dr. Shanker was in this state, alone. . . . All his fears for Yariko came rushing back.

Shanker stared at him a moment with a confused expression. Then he shook his head. "Not here. I don't know." His voice was so low that it was difficult to understand.

"You left *her*?" Julian said, tightening his grip on Shanker's arm. He felt no sympathy at that moment; only anger that he had left Yariko alone.

"You're a fierce young fellow," Shanker said, summoning up the energy to speak again. "I don't know where Hilda is either. I hope they stayed together." He closed his eyes to rest, but a moment later he looked up again and said, in a weak voice, "Who's your friend? Doesn't he speak?"

"Yes, English."

"You're a marvel," he mumbled. "You've already taught the natives English."

Julian glanced around but Carl had disappeared. By now he was used to these sudden absences. Carl often scouted ahead, or climbed an escarpment to spy out the land. This time their guide came back within a few minutes, carrying a skinful of cold water from a nearby stream. The water helped to revive Dr. Shanker and he looked up at them with a more focused attention, droplets sparkling in his beard.

"Strange native fauna. . . . I must be hallucinating," he mumbled, staring at Carl and blinking.

"But what happened?" Julian said, stooping down beside him. "How did you lose Yariko? Is she still alive?"

"Can't you think of anything else?" Shanker said. He looked

haggard but his voice was stronger, already taking on its usual egotistical tone. He stared at Carl again, then closed his eyes and rested a moment. "Been having funny deliriums for days . . . but this one beats all."

Julian didn't have any patience with the man's confusion. "Tell me what happened," he said again.

"I'll tell you," Shanker said at last, "just so you don't pester me about it. The river . . . you fell in the river. I thought you were done for. We couldn't do much for you, being occupied by our horned friend. Standing right next to one of them for the first time—a live one—is something I won't forget."

Julian mumbled in agreement.

"Of course we didn't know what was happening to you, except that you didn't join us in facing the thing. I don't think the animal was trying to kill us, or it would have. We couldn't have fought it. There was a young one behind it; maybe it was a straggler and the parent had come back for it."

Julian nodded again.

Dr. Shanker took another sip of water and continued. "Anyhow, Mamma was warning us off in no uncertain terms. She came out to the river bank. That's about when we realized you'd gone for a swim. Not bright, if you ask me; that water was cold. Yorko was frantic. She lost it, for moment: she actually charged the thing. Straight along the bank, screaming bloody murder. She practically impaled herself on its horns. That almost ended things: Mrs. Triceratops decided to charge us. Chased us quite a way upstream, in fact, snorting and squealing like a monstrous pig. She seemed unimpressed by our spears. I wonder if even a T. rex can bring down one of those. Quite terrifying."

"But later, what happened later? Why didn't we find each other?" Julian still didn't have much room for sympathy.

"We searched all afternoon, Whitney. Finally picked up your trail where you left the river. But . . . maybe it wasn't even your trail," he said thoughtfully, looking at Carl.

"You were right behind me, then," Julian said. "I thought I was following you, but you were following me. No wonder I couldn't catch up to you."

"Comic," Dr. Shanker said. "Typical. I knew we should have gone sooner, but Yorko wouldn't listen." He shrugged. "We finally

224

moved on, but we didn't get very far. I can't say I like this rocky ground any better than the swamp. Cluttered up with too many of those damn Triceratops."

"They're migrating," Carl said, quietly.

"What's that? What did he say?"

Julian repeated the word; he remembered that Carl's speech had been a little strange to him too at first.

"That's his hypothesis?" Dr. Shanker said, peering at Carl with his bushy eyebrows drawn together in a scowl. Carl said nothing, but Dr. Shanker continued to stare. He was obviously as amazed as Julian had been at the first sight of another human.

"But where's Yariko," Julian said again.

"All right, all right. Where was I? That second night, in the middle of that God-awful tempest, I stepped on a diabolical little reptile. It bit me on the ankle. I killed it with the back end of my spear, broke its spine, which is some consolation, but the rotten thing caused my ankle to swell up. I ate it the next day."

"What did the reptile look like?" Carl said, glancing at them over his shoulder. He was squatting on the ground rummaging in one of his sacks.

"What did the what? Reptile? Whitney, what did he say?"

"The reptile," Julian repeated. "What did it look like." He wondered if there were poisonous lizards in the Cretaceous.

"A little thing," Dr. Shanker said. "A foot long at most. It was ringed with red, I think, but it was hard to tell in the moonlight. Red and yellow. What's the verdict?"

But Carl said nothing.

"Queer fellow," Dr. Shanker said. "Anyhow, that's when we got separated. Yariko disappeared in the rain—it was pitch black by then and I couldn't find her. Then the dry stream we were in flooded. There'd been water on the ground all night but suddenly it started moving. It knocked me down. I don't know about your river experience, Whitney, but mine wasn't exactly fun." Shanker scratched at his hairy face and coughed.

"Was Yariko washed away too?" The thought of Yariko carried away by a river as he had been was a terrible one.

"I don't know, but I don't think so. The water wasn't that deep. I'm sure we would have ended up in the same place, or close enough. I lost Hilda, too. When I managed to crawl out I couldn't walk very well though I tried. I circled the area calling. No sign of her in the morning, or Hilda either. Fantastic place for lightning, by the way, this prehistoric period of yours. You never warned us. And then the fog came, so it was all hopeless. I've been staggering along ever since, trying to catch up. You see, I've found bits of Hilda's fur and a few dog prints here and there. I'm sure they've come this way."

"Then she must be nearby," Julian said eagerly.

"I doubt that," Dr. Shanker said dryly. "She was so anxious to catch up with you that she must be miles ahead by now. I haven't exactly been traveling at my peak speed." He grinned, but the grin quickly turned to a grimace. "This damn foot of mine." He closed his eyes and leaned his head back to rest.

Julian looked at him as he sat back against the stone. He did not look wasted; his frame was still burly. But even through the hair and the filth on his face the unhealthy glow of a fever could be seen. The wet and exposure, lack of good food, and constant physical strain must have weakened him. No wonder he was in such bad shape. Worried as he was about Yariko, Julian felt ashamed of his impatience toward a man who was so sick.

Dr. Shanker opened his eyes and looked up. "Whitney. Leave me some food and water, and I'll be fine here. I know you want to hurry after her. I'll catch up in a few days."

Julian glanced at Carl, who gave a slight shake of his head. He had lit a fire and was warming some meat and roots on the stones. He sprinkled the meat with a crumbling of dried leaves, and the spice improved immensely on the flavor. Even Dr. Shanker mumbled something about it as he gnawed on a piece.

After eating they took off Dr. Shanker's shoe and carefully washed his ankle with relatively clean water. It was badly swollen and streaks of red and black extended up his leg, under the skin. The cold water seemed to lessen the pain. His pant leg was cut off a few inches short; obviously the scrap of cloth Carl had found was an attempt at a bandage.

Carl rummaged in his sack and pulled out some dried stems, twisted and hard. Julian could tell nothing about the plant except that it was a monocot. It had a pungent smell, something like pepper or cinnamon, too strong to be pleasant.

Carl broke off a piece and handed it to Dr. Shanker. "Tell your friend to chew it," he said to Julian. "It will make the swelling go down."

"You don't need to interpret," Dr. Shanker muttered. "I can understand him well enough now." He took the twig and sniffed at it, grimacing. "Pleasant stuff, your voodoo medicine." But he chewed it nonetheless. "Salicylic acid. Aspirin. In Boy Scouts we knew a dozen plants that contained the stuff."

Soon after, he lay down and went to sleep in the small shadow of the bushes.

Julian sat beside Carl and looked out at the bushy canyon. The sunlight glared and sparkled from the stones and the pools of water. A cloud passed overhead, dimming the light. Then the cloud moved on and the light glared in his eyes again, as fiercely as ever.

For a long time they said nothing. Julian was learning his guide's habit of silence, of conserving words as if they were a finite resource. Finally he said, "Will he get better, do you think?"

"If the black lines fade by morning, he will be well enough to travel. If they grow longer, he will die."

"Do you think that'll happen?"

Carl shrugged, and said nothing more.

Julian guessed that Dr. Shanker was suffering from a blood-born infection. The modern Komodo dragon used the technique of infecting its prey with bacteria harbored in the serrations of its teeth, and it seemed plausible that some of the Cretaceous carnivores might follow the same strategy. Ironically, the reptile had not only lost its meal, but gotten itself eaten for its pains.

After a while Carl reached into the stiffened sack and took out a strangely shaped piece of wood, flattened on one side and glossy as though it had been polished with oil. It was about the width of his spread hand and maybe four times the length, and pegged to the flat top were five white strings of decreasing length, gut perhaps, thin as thread and pulled tightly over the wood. Beneath the strings, at one end, a hollow had been carved into the wood, about twice the size of a fist.

Carl set the contraption upright against his chest, the hollowed end resting on his crossed legs, and plucked at it with his fingers. His movements were graceful, despite the missing fingers on his left hand. Because the strings were so short and the sound cavity so small, the instrument had a pingy sound with little resonance.

The music was strange and quiet; it reminded Julian of Yariko's Japanese instrument that he'd once heard her play, so long ago. He leaned his head against the rock and listened to the soft wandering sounds that brought Yariko so clearly to mind.

Finally Carl said, without looking up, "Lie down and sleep. Tonight we have a hard journey."

Julian looked carefully at Dr. Shanker. The man seemed to be sleeping peacefully, probably for the first time in days. Rolling up his poncho, he took Shanker's swollen foot and propped it up. Then he put the water skin near the sick man, and stretched out on the ground beside him. "And you?" he asked, looking at Carl.

"The young need sleep," Carl said, quietly. "I do not."

Julian woke to find the sun past its height and Carl gone. He stood up to stretch and saw Dr. Shanker walking about in the sunlight, limping heavily and leaning on his spear. He looked pale but his face was set in determination, and the fact that he was standing at all was a big improvement.

He nodded to Julian. "Fantastic day," he said, and it was. The sun gleamed on the rocks and a few birds fluttered in the bushes on the other side of the stream bed, calling to each other in short, raucous notes. The signs of life were heartening after the silence of the past few days.

"I'm glad to see you up and striding around," Julian said. "Carl thought you might . . . be sick for quite a while."

"I've been up for at least an hour." Dr. Shanker gestured at a small pile of twigs and twisted wood that he'd gathered. "Your friend's disappeared, anyway," he added.

"He comes and goes. He's probably scouting, or hunting."

"Strange fish," Shanker said. "Where did you get him, Whitney? Don't keep me in suspense. Is he for real? How did he get here?"

"I wish I knew." Julian sat down and removed his shoes, or what was left of them, and picked the grit out from between his toes. "As far as I can understand, a group of people got here before us and survived for a couple of generations. Carl's the last one. One of the grandchildren or even great-grandchildren, I gather; he never talks about coming from the future. Actually, he doesn't really talk at all. I think he's been alone for a long time, maybe decades. I wish I could have a bath," he added, plucking at his sticky shirt.

"You think it was a separate translocation event? An independent group of people? Not possible." Dr. Shanker lowered himself carefully onto a boulder. "Damn this foot," he grumbled. "Think about it, Whitney. The history of the earth, as a distinct gravitational entity, is about six billion years. We go back in time and hit the Late Cretaceous. Somebody else goes back in time, and what's the chance that they'll hit within a hundred years of us? One in six hundred million. Negligible. It can't happen. On the other hand, here we have your stalwart friend. Maybe we're both crazy and imagining him together. Are you sure falling in that freezing water didn't addle your brains? Maybe you banged your head on a rock."

Julian grinned at that. "And you—brains addled by a lizard? Anyway, I'd already figured all that. Obviously they got here the same way, with the same equipment. Remember we used to joke about the rescue party? Maybe there really was one, or an attempt at one."

Dr. Shanker grunted. "Who would know enough to set everything up again and get in the vault during a run? And if they did know that much, why would they risk themselves? No, this translocation through time isn't something anyone's going to figure out, unless we get back and tell them." He pondered for a moment. "Didn't you ask him about his origins? He must know something."

"Of course I asked him," Julian said, realizing as he spoke that he hadn't asked all that much after that first day with Carl. Perhaps his amazement had been blunted by Carl's familiarity and the natural way he seemed to fit into the landscape; and his curiosity had been buried under his desire to find Yariko and Dr. Shanker.

"He talked about people coming from the east, from the edge of the sea. They traveled west: so maybe they did know what they were doing, if they were trying to revert." Julian eased his shoes back on over the blisters. "He's referred to maybe several generations, but I can't be sure. I told you, he isn't much of a talker. In his own good time, I suppose."

"Your lack of curiosity astounds me, Whitney. And here you are a scientist, a Cretaceous expert at that. You'd never let a new dinosaur off that easily. I intend to get answers from the guy. His horrible voodoo medicine was just what I needed, anyway. I'll never underestimate aspirin again."

"Yes, he's remarkable. He saved me from Corla."

"Corla?" Dr. Shanker peered at Julian from under his bushy

brows. "You've led a whole life, I see, since the last time I saw you. You'll have to tell me about it."

Perhaps he was right, Julian thought. It has been like a new life.

Dr. Shanker nodded at the gully. "Here he comes."

Carl was approaching with a bulging sack slung over one shoulder. He stopped in front of the others and put down the sack without a word. Dr. Shanker winked at Julian and said in a low voice, "You're right. Not a talker."

Thanks to Dr. Shanker's kindling, they ate a warm meal of fresh tangy roots and stems, a nice change from the dried meat, and drank from a stream that trickled down the wall of the gully. Then they set out. Carl took the lead; Dr. Shanker came next, crutching along at a surprising pace with his spear, trying to gain the lead but not succeeding. Julian came last of all, trying to contain his impatience. He was eager to find Yariko again, although he supposed Dr. Shanker was just as eager to find Hilda.

As Carl had warned them, the terrain was rough and cut by gullies and sudden clefts. Shrubs and gnarled pines dotted the area, but did not give much shade. The sun burned down on them. Julian's hair was noticeably hot to the touch and his back, underneath the water sack, was soon drenched in sweat and chafed by the leather. They stopped now and then to drink from a brook or a trickling waterfall; Dr. Shanker bathed his foot each time, sighing as the cool water ran over it. The swelling, to Julian's relief, was going down.

Carl did not show any fatigue but as the day went on Dr. Shanker lagged and his limping grew worse. He never complained, but kept pegging away with the butt end of his spear. For all his arrogance, Julian had to allow him a certain tenacity and consistency of character. The man seemed completely unfeeling about his own condition. Julian wondered when he would begin questioning Carl.

After several hours of hard traveling they began to climb steeply uphill. At the top of a ridge, panting and leaning on their spears, they looked to the west and saw the whole immense landscape spread out like a map. The broken ground continued, stained brown and green and laced with canyons and gullies. In the distance the ground rose again, easing up toward a group of isolated hills.

To the best that Julian could calculate, those hills marked the place they had been trying to find, the end of their thousand-mile journey. He hoped that Yariko at least would reach them in safety. The region, open and burning in the sun, seemed to hold fewer carnivores and might be less dangerous. But the hunters were scarce precisely because their prey was scarce. There was obviously little food to be had, and Yariko probably did not have a supply slung on her back, as Julian and his companions did. And Dr. Shanker's injury was a reminder that large carnivores were not the only things to fear.

At some point the old river bed they'd been following fanned out and disappeared in the rocky landscape. Julian never expected to find a trail nicely laid out all the way to their destination; still, he was disturbed to see it disappearing, because their path would probably begin to diverge from Yariko's.

They rested while the sun went down and then continued in the dark under the moon, now nearing the third quarter. More gullies, ridges, broken ground; a bleak and barren landscape. Julian saw no animals, and no signs of Yariko either. Another cliff face loomed ahead, and they stopped in its shadow to give Dr. Shanker a rest. Progress had been good; but reaching their goal within the limited time window would be meaningless to Julian if they did not find Yariko.

As he sat there, trying to make himself comfortable on the hard stones, Julian thought that he felt the faintest trembling of the ground.

Carl seemed to have caught it too. He rose. "We need shelter. We have no escape here in the open."

"Escape from what?" Dr. Shanker asked, looking at Julian. "I thought you said there'd be fewer of those pesky large animals out of the swamp, Whitney." He pulled himself to his feet with the help of his spear. "What do you see, you two?"

"She is following us," Carl said.

"Who's he talking about?" Dr. Shanker said. "A pet dinosaur?"

"I have fed her over the years," Carl said, "and now she expects it."

"Corla?" Julian asked, amazed, and Carl nodded.

Julian wondered if *T. rex* had that kind of memory. Tyrannosaurids had immense heads, but they were mostly bone and muscle. The braincase was relatively tiny. Still, she might have learned to associate humans with a steady supply of food; and she could certainly trail them by scent.

Carl led them toward a thicket of pine trees that crowded the base of the cliff. He said the strong scent of the needles would help to hide them. They settled in the darkness under the trees, backs up against the shaggy trunks, and ate a little dried food.

"What is it that's following us?" Dr. Shanker asked.

"One of the Big Ones," Carl said, as if that made everything clear.

"A friend of yours?"

Carl said nothing. He continued to cut off plugs of meat with his bone knife and chew them.

"Our wise old guide refuses to speak," Dr. Shanker said, irritably. "What does he know about Cretaceous animals anyway? How can he be sure this thing won't crash over here and attack us? Whitney, maybe he'll answer you. He seems to like you."

Carl looked at him for a moment, his eyes narrowed. But there was no anger in his expression; instead he seemed to be weighing his opinion of Dr. Shanker. Finally he thrust a plug of meat into his cheek and said, "I am sure of nothing."

"It's a *T. rex*," Julian explained. "She must have followed our scent from Carl's hill—from near the river."

With his mouth still full of chewed food, Dr. Shanker laughed, a little uncertainly. "Is that plausible? Do they even have such a well-developed sense of smell?"

"They're known to have enormous turbinal bones, the spiral bones in the nose," Julian explained. "Her sense of smell, I'd guess, is probably uncanny with a surface area like that." He didn't mention the other thing he was thinking: Corla was used to being fed by this man; she had followed him into the barren hills, far from the river valley, where there was no prey to be caught. Carl had nothing to give her. Dr. Shanker interrupted his thoughts.

"And you imagine that this animal is specifically tracking us? Up hill and down? Brainy, for a dinosaur. Ph.D., perhaps?" He snorted. "I think you're giving it too much credit. It isn't like Hilda, you know. We seem to have thrown it off our trail, anyway. I don't hear anyone stomping around out there."

Carl seemed impervious to the sarcasm; he may not have understood it. He chewed, spat out the end of the meat that had become too dry to swallow, and said, "Would you like to see her?"

Dr. Shanker stared at him, then stared wildly around at the trees. "What, here? It's that close?"

Carl nodded, continuing to eat.

Julian was also surprised; horrified, in fact.

Dr. Shanker carefully set down his piece of meat on a rock. "Then what are we doing, sitting here waiting to be killed?" For the first time he spoke directly to Carl.

"Resting. Safer here than in the open."

Dr. Shanker scratched his beard and continued to stare at Carl. Julian could not tell through all the hair on the man's face whether he was angry, or frightened. After a moment Shanker said, as if to himself, "I'd hate to leave here without ever having seen one . . . that is, if it doesn't involve getting killed."

"Then be silent, and follow me," Carl said, rising.

The proverb of putting all your eggs in one basket could apply to putting too many dinosaurs in one mold. Dinosaurs became increasingly specialized, and by the end of the Cretaceous were restricted to a few ecological niches, primarily large animal niches. Hadrosaurs, for example, are known for their wonderful variety of size and shape, and especially for the range of fantastic crests on their heads. By the late Maastrichtian, however, the predominant hadrosaur was Edmontosaurus, *uncrested, gigantic. Likewise,* Triceratops, *the largest of the ceratopsians, became predominant, while other ceratopsians disappeared. This loss of diversity almost certainly made them more vulnerable to extinction.*
—*Julian Whitney*, Lectures on Cretaceous Ecology

— TWENTY —

2 September
12:16 AM Local Time

Earles closed her door again and turned back to Mark. "What something else?"

He handed her the piece of paper.

It was a printout of a digital picture; a picture of a garishly colored beetle with very long antennae and spiky legs. Earles looked up. "This is one of their mysterious beetles? The ones that appeared, and then disappeared?"

"Yes," Mark said. "I found the image file and printed it. Those scientists you brought in—they won't let me near the main computer. I had to wait until they took a break."

"You think there's something special about this insect," Earles said, handing back the picture. "What's the book?"

Mark held it out; it was a field guide to North American insects. "I found this in the lab. It belongs to that paleontologist they called. His name's inside it."

"And you've looked up the beetle?" Earles' mind was racing; if this insect could be localized to a particular region, authorities there could be alerted to look for the missing people . . . unless it was a very remote region. Something about the beetle's color

234

made her think of tropical jungles.

"This beetle isn't in the book," Mark said. "And why would they call a paleontologist? Why not that entomologist in the ecology department?"

"Tell me what the difference is," Earles said.

"An entomologist studies insects," Mark replied. "Usually modern insects. A paleontologist studies things of the past—fossils and such. Things that don't exist anymore."

Earles wasn't sure she was following him. "Those scientist thought they were recreating extinct beetles? You think that's why they called in that guy Whitney?"

"Why else would they call him? I've never heard of a physics lab bringing in someone like that. I showed the picture to the entomologist, Bob Heckwood. He couldn't come close to identifying it." Mark looked straight at her for the first time. "I think they were retrieving samples from sometime in the past. Like, millions of years in the past. I looked up that guy Whitney's profile. He's an expert in Cretaceous ecology. He wrote a paper on Late Cretaceous beetles, in fact."

"An expert in what?" Earles began to think she was taking a crash course in scientific terminology.

"The Cretaceous Period. You know, dinosaurs and all that. It was just before the dinosaurs went extinct."

Earles nearly let her jaw drop but managed to control such an undignified response. "Are you suggesting that those missing people. . . ."

"Were sent back to prehistoric times, maybe the time of the dinosaurs. Yeah."

• • • • •

Julian stared at Dr. Shanker as if seeing him for the first time. "You're out of your mind!" he hissed.

Dr. Shanker grinned. "Whitney, I'm surprised at you. And when you're back home sitting in your office, sipping your coffee, you'll never forgive yourself for passing up a chance like this."

Julian mumbled something about a chance to get swallowed whole. He had already stumbled on Corla one night and been very nearly killed. In retrospect, he knew the attack had been less than halfhearted, or he would not have survived. Maybe she had only been curious that time. But even the idle curiosity of a *T. rex* did not much appeal to him.

Still, Julian hesitated only a moment before following his companions out of the trees and across the scrubby plain to a black mass of tall, tangled bushes. At the edge of the thicket Carl stopped, studied the ground, and paced silently back and forth. Then by gestures he indicated a huge gap in the hedge, leading into the darkness under the forest canopy. Something had already come that way and broken through the tangles of vegetation: something enormous.

Carl leaned in close to Julian and said in the barest of whispers, "Listen to her breathing."

Julian had been listening to it for a while, thinking it was the wind in the branches. Now he realized that it was regular, and immense.

Peering into the rent in the foliage, he gradually made out blotches of light where the moon filtered down through the canopy. He could see a rounded object. The large theropods were thought to have crouched down on their bellies when they rested, if they slept at all. Julian assumed, therefore, that he was looking at the curve of Corla's spine. Enough was visible to give some sense of the scale of the animal. At eighteen tons, she would be two hundred times the size of a human. Julian began to get the shakes and wished like anything they had never come so close. But at the same time he was absurdly pleased that they'd succeeded in creeping up on the thing; stalking the stalker, this ultimate of predators.

At a signal from Carl they turned and crept away. Carl did not follow immediately, but let the other two get some way ahead. Julian wondered if he was trying to protect them: if the tyrannosaur woke up, he would be the closest to her. Looking back, he saw Carl still standing there, staring into the darkness at his long-time friend. That friend, if she saw him, would probably kill him.

Dr. Shanker and Julian had crossed the open, moonlit area and were near the shelter of the pine trees before they stopped to look back again. Carl had just turned to follow when behind him the monstrous head of Corla emerged out of the shadows and bobbed, as if sniffing the wind. Julian froze and stared; but Carl made a gesture for them to continue toward the trees.

"Get under cover," Dr. Shanker said in a low voice. "There's nothing we can do for him, if that thing spots him."

"We can't leave him," Julian whispered back. He watched as Carl strode across the open area, through the rocks and low shrubs. The man's life depended on reaching cover before the animal saw him; but if he ran, he would certainly attract its attention. With no hand-out of meat forthcoming, would the animal take Carl instead?

Carl gestured again, and reluctantly Julian led the way into the shadows under the trees. But as soon as he was hidden from view, out of the direct moonlight, he stopped and looked back again. Carl was still walking toward them, carefully, silently. Well behind him, but still towering over him, paced the tyrannosaur. How, Julian thought to himself, can a thing so monstrous move with no sound?

He picked up a small rock, hefted it, and stepped out of the fringes of the wood. Carl, seeing what he was doing, gave a slight shake of his head.

"Does he know it's there?" Dr. Shanker whispered from the shadows behind a tree.

"Of course," Julian said.

They watched the scene progress, silent in the moonlight: the man in front moving with long, deliberate strides, and behind him the huge beast stepping with strides five times as long. But she did not seem hurried; she kept an even distance, as if she intended only to keep him under observation.

"Like a dog following its master," Dr. Shanker said.

The animal stopped when she reached the center of the open area. Her shadow in the moonlight stretched huge over the stony ground. She stood perfectly still, alert, watching Carl as he approached the trees. The stump of her clawless forelimb gave her a lopsided, almost pitiable appearance; but in reality it made little difference to her weaponry.

"She has eaten," Carl said, as he came up to them. "Do not provoke her."

They turned and crept into the shadows of the trees. For a long time nobody spoke. The moon set, and real darkness came on. Julian waited for Dr. Shanker to crack a joke, but instead he was silent. He seemed uncharacteristically subdued. After a long time Shanker said, quietly, "You're a brave fellow, Carl."

They continued into the forest, following the line of the cliff. When Julian glanced back for the last time he could see her still standing there, filling the center of the stony clearing, gazing at the place where they had disappeared.

The next morning the travelers increased their pace and rested only occasionally. All day there was no sign of Corla; Julian and Dr. Shanker had hopes that she'd lost their trail or wandered back to her home territory. Carl said nothing.

In the mid afternoon they came upon a circle of char, a few burnt bones scattered on the ground, and at the very edge of the circle, pressed into the soft ash, the clover-shaped footprint of a dog. Julian was ecstatic. It was the first indication that Yariko was still alive, and that Hilda was with her. He felt immensely better knowing Yariko wasn't entirely alone. But Carl sniffed at the ashes and said they were at least a day old.

Far ahead, just visible on the horizon, a distinctive peak jutted out of a group of low hills. Carl said they would head for it. Julian nodded; it was the only obvious landmark in sight, and it would make a good base to search for Yariko. She might even make for the same point herself. It was probably an old volcano; but it could not have erupted in the past many years, judging by the patches of green clinging to the sides and the top.

The shadows lengthened. Carl speared a small dinosaur, a bipedal thing only two feet high. They cooked it when they stopped to rest.

"What, not prying its mouth open to see the teeth?" Dr. Shanker said as they chewed bits of the singed hind legs. "Feeling all right, Whitney?"

Julian looked down at the head and other less appetizing parts of the dismembered animal. He had never seen a skeleton of this type before; but strangely, the teeth seemed of remote importance. Nevertheless, he poked at the mouth with a stick and glanced at the dentition. "I'm not familiar with this pattern," he said, and then dropped his stick and stood. "Shouldn't we get moving again?"

Carl stamped out the tiny fire. "The land will become rougher as we go," he said with a glance at Dr. Shanker.

The information was disheartening. More and more they'd been forced to make detours around sudden canyons or slopes of scree that made dangerous footing. After the smooth journey up the ancient river bed for so many miles, Julian felt as though they were crawling along, barely making progress.

"How far do you think we have to go still?" Dr. Shanker whispered to Julian as they started out. "A hundred miles?"

"Seven days' walk." Carl spoke from up ahead.

Shanker stopped, causing Julian to stop behind him. "I was asking you, not him," he said quietly, indicating Carl with his head. "We should talk more about where he's taking us; make sure it's the right direction."

"We're going west, and straight toward an obvious landmark, anyway," Julian answered, in a normal voice. "It seems the likeliest way to catch up with Yariko. Besides, if there are any signs of her, he'll find them. We'd probably miss them."

"All right, Whitney, if you're sure. We do seem to be going the logical way—for now." Shanker hurried after Carl.

Julian followed thoughtfully, pondering the terrain and what they might encounter. The high ground near the mountains was entirely unknown to modern paleontologists. The soil now under his feet was already eroding, washed away by mountain streams and weathered by the wind that blew down from the west. Fossilization under such conditions was nearly impossible. Sixty-five million years in the future, his colleagues would have little idea what types of plants or animals inhabited such dry, elevated areas. Everything Julian saw, however solid, bright green, or full of life, would disappear within a million years or so and leave no trace of its existence.

They walked perhaps another six hours, until the night was completely black. The moon was waning, rising later, providing light only in the second half of the night. There was one small alarm: Dr. Shanker was nipped on the toe of his shoe by a tiny creature. "Do I have tasty feet, or something?" he roared in exasperation as he stumbled. "Here, Whitney—I've got it on my spear. See what it is."

When they stopped once more, Julian was glad of another fresh meal. He called their dinner *Impalidus shankerensus*. Carl warned that it might be their last fresh meat for some time; Julian grimaced at the thought of the dried, tasteless strips of meat that had been collecting dirt in the bottoms of their sacks.

They crouched around the fire. It was cooler at night out here than they were used to. Dr. Shanker inspected his ankle in the flickering light. "Excellent," he said. "Healthy. Not absolutely gangrenous, anyway." The truth lay somewhere in between; but he'd been walking more easily and the swelling was almost gone. "How much more of this scrambling over the rocks?" he asked.

Carl squinted into the fire. "The full seven days. The path will be difficult near the mountains."

"Charming. How often do you come this way?"

"When I was younger, every year. But not for many years now."

"What about rockslides, volcanoes, earthquakes? How do you know that the path is still there?"

"I do not."

Dr. Shanker nodded. "That's honest enough. Tell me, where did these people of yours live?"

Carl pointed a little north of west. "In the caves."

"Where did they come from?"

"Those before came from the water in the east, as you did."

"I told you all that already," Julian put in.

"And what brought them all this way to the mountains?" Dr. Shanker ignored Julian.

"The same thing that brings you."

Julian looked up; he'd never put the question just that way.

"And what was that?" asked Dr. Shanker, after a moment.

"You know; I do not," Carl said shortly, and lying back on the ground, he closed his eyes. He had had enough of being questioned.

There was little rest that night. Carl soon rose and led them on, under the crescent moon. The shadows of trees stretched black and crooked across the ground. Every time a branch stirred in the evening breeze its shadow would skitter over the stones like a pack of animals. Julian could hardly see Carl ahead, blending into the patchwork of black and gray, and even Dr. Shanker's form directly in front took on the appearance of a bear, lumbering awkwardly. The strain of the hike was beginning to tell on Julian. He found himself jumping and starting at every sound and every movement.

"If we still had the gun. . . ," Dr. Shanker muttered. His voice seemed to disappear into the blackness. Julian wished he would keep silent; he did not want any predators taking notice. "But I suppose," Shanker went on, "our little piece would have been a rum weapon, to fight off a T. rex. Ever use a gun, Carl?"

"What is a gun?" Carl's voice floated out of the darkness up ahead.

"It makes a loud noise. It throws a piece of metal fast enough to go through any dinosaur."

"Noise we do not need," Carl said.

For several days the three scrambled over the rocks, crawled through thickets of bushes, climbed down into gullies and out the other side. Sometimes the gullies were filled with lush vegetation taking root in the accumulated silt and mud. They used their spears to beat a way through these tangles of ferns and leaves. In the darkest part of the night just before dawn, when even Carl could not see enough to continue, they huddled under a thicket of bushes or trees and tried to keep warm.

In another age the travelers could have bowled over a big shaggy animal and taken its pelt; but they had to make do shivering under the leathery, hairless dinosaur ponchos. Julian knew it was not actually all that cold; but their time spent in the steamy swamplands made them sensitive to the dryer, cooler air of the hills. He found himself dreaming wistfully of giant sloths and wooly rhinoceroses.

Their food supply dwindled. Carl explained that the small prey animals usually stayed hidden in the dense foliage of the gullies, so they spent most of one day following a canyon to the north, far out of their direction, trying unsuccessfully to flush out dinner. In the evening they retraced their steps, but leaving the gully too soon became lost in a field of jumbled rock and thorn bushes. The sun went down; the light was too dim to make out the peak they were steering for. Carl called a halt until daylight.

It was a cold, shivery night. Julian's stomach felt pinched; there'd been no food in nearly twenty-four hours. Dr. Shanker, to Julian's envy, slept perfectly soundly. As usual, Carl disappeared for most of the night; he seemed to survive on almost no sleep. When Julian woke partway through the night, he saw Carl stretched out on the ground beside him, staring peacefully up at the stars.

When they woke Carl said he'd found a path that was relatively clear. It ran beside a low cliff that swept downward from the hills. Since it maintained a fairly straight westward course, their best option was to follow it. He had also found the tracks of a small herd of animals.

Dr. Shanker sat nearby, sharpening his spear on a stone. "I intend to kill one of them," he announced. "Whatever they are. Even

Triceratops." He jabbed the point into the soft ground. "What do you think?"

"You cannot kill the Horned Ones with a spear," Carl said.

They set out, single file as always.

"Was it Triceratops prints that you saw?" Dr. Shanker asked.

"No," Carl said. "They do not live here."

"What was it, then? Big? Small? Do I spear it from behind or in front?"

"You decide," Carl said. He pointed up the steep cliff.

Julian looked up at the gray lumps of rocks and the green of bushes tufting out of cracks here and there. Then something moved, and a few pebbles rattled down the cliff face. The animal was so much the color of soil and bushes that it was superbly camouflaged as long as it remained still.

"Aha," Dr. Shanker said, very quietly. He began to creep closer.

It was a quadrupedal animal in the ceratopsian family, but it was quite small, about the size of a large dog. It watched the humans, cautious but not yet afraid; a few bits of shrub protruded from its beak-like mouth.

Then Dr. Shanker rifled his spear. The cast was impressive, but missed anyway. The point bounced off of a stone only a few feet away from the animal. Instantly, the creature scrambled away up the steep incline.

At the same time the whole cliff came to life. At least ten other animals leaped into motion. They bounded up the cliff and disappeared over the top, grunting and blowing. One of them, however, lost its footing and slid backward several yards. It seemed on the edge of regaining its balance; but then in a cascade of pebbles and dust, it plunged down the cliff, struck hard against several boulders along the way, and landed very nearly at Julian's feet.

"I told you I'd get one," Dr. Shanker said, grinning.

He dispatched the dying animal and they dragged the carcass into the shade of some nearby trees. It was so badly cut from the stones that it left a wide, bloody trail through the weeds and ferns. The whole place reeked of blood, and Dr. Shanker joked that if Hilda were nearby she would be sure to come running.

"I hope nothing else comes running," Julian said.

Carl said nothing. He had already started to skin the animal with his bone knife. While he and Dr. Shanker dressed the carcass, Julian gathered twigs and sticks for a fire. If they could smoke the meat they'd have enough for many days.

Just as he finished arranging the kindling into a neat pyramid, Dr. Shanker looked up from his carving and said, "We have a dinner guest."

Humans have a tendency to project their own rationalities onto other animals: anthropomorphizing, ascribing motives that do not exist. Thus someone will say, "My cat threw up on the rug because he was mad at me," when the cat had a very different, and far more sensible motive: to get something dangerous out of his digestive system. So it is with any wild animal. Each has its own particular needs and motives, aimed at personal survival and procreation of the species. So it was with animals now extinct; as paleontological ecologists, we must ascribe Cretaceous motives to Cretaceous animals.
—*Julian Whitney*, Lectures on Cretaceous Ecology

– TWENTY-ONE –

2 September
12:52 AM Local Time

The route from the station in town to the physics building on campus was becoming far too familiar, Earles thought to herself as she accompanied Mark. He had left the beetle picture and book in her office, and agreed to go right back to work on the wiring in the graviton vault. He could also install the platinum bar in the circuit he was rebuilding.

Earles planned in her mind what she would say to those two physicists. Who did they think they were, trying to deceive her and steal valuable scientific information? Wanting to leave other people stranded somewhere unknown (surely Mark was overdoing the guesswork by suggesting they were off in dinosaur times), perhaps in danger, maybe unable to return on their own; the whole thing was crazy.

She realized Mark was having to jump every few steps to keep up with her. Slowing her stride, she went back to her thoughts.

Of course, they might not have any such intent. Perhaps they simply didn't know; maybe they believed they were doing the right thing. But the more she thought over the last conversation in the lab, when Ridzgy had cut Bowman off and handed her the measurements—thus effectively sending her away with a project to keep her busy—the more convinced Earles was that something was being kept from her.

It was a delicate situation, and must be handled carefully. No charging in swinging accusations. That would get her nowhere. She knew a better way. She slipped her hand into her pocket and felt for the cold bit of metal.

"You go in first," she said to Mark when they entered the physics building. "Go back to work and don't say anything. No need for them to know you've spoken to me, even if they are innocent."

"They're not innocent," Mark said, but he went on down the grimy stairs to the basement lab.

Earles paced the empty, silent lobby for five minutes, thinking. Then she followed Mark.

• • • • •

He was already in the vault when she entered the main room; she could just see his feet on the ladder and hear a scraping sound that must be wires pulled through conduits.

The two scientists looked startled to see her, as well they might be. It was one in the morning. Bowman's pouchy face was gray and stubbly, his eyes baggy; he looked a bit the worse for wear. Ridzgy was clearly tired but also clearly determined.

"Still can't stay away?" Bowman said with a false cheeriness.

"I've brought the platinum bar," Earles said, placing the small box carefully on the bench. "Where is the original, by the way?" She looked around the counter and spotted it, sitting on an open notebook like a paperweight. Somehow the casual use of this critical thing annoyed her. She picked it up and studied it. The metal looked identical to the replacement except for one end, which was molded like silly putty into a smooth, drooping mass.

"Well, good thing we were able to replace it," she said lightly, putting it back on the notebook. "How long until the program is ready to run?"

"Soon," Ridzgy answered. "There were several versions, so we'd like to dry-run them all, you know, without the actual vault, and see what numbers come up. Then we can be sure of running the correct version when the time comes. Once that kid finishes the wiring we can install this platinum."

"I think Mark is planning on doing that step, since it involves the electronic circuit. How's he coming along?"

Bowman shrugged. "Haven't checked. Seems to be working hard. Doesn't say much to us."

Over Bowman's shoulder Earles saw Mark peering out of the vault; once again, he looked confused. And no wonder, if he'd overheard the conversation. He climbed out and approached with a determined look on his face.

Earles gave an almost imperceptible shake of her head and he stopped. "What progress?" she asked, and the two scientists turned to look at Mark.

"Um, going well," he said, still staring at Earles. "There are some things that I need, small things, from the electronics storeroom. I'll have to track down campus security to get in though."

"Well, get on it," Earles snapped. "And you two: go with him. I want campus security to know who you are. Should've brought you together hours ago, really. It wouldn't do for some guard to detain you half the night as trespassers." When nobody moved she roared, "Well?"

Mark scampered out of the lab, and the two physicists followed more slowly.

Earles listened for the outer door being closed. Then she went quickly to work. A small recording device, no bigger than a wallet and holding a tiny disc, was easy to conceal; she put it inside one of the filthy coffee cups and covered it with a crumpled paper towel. The cup was in an inconspicuous place and hadn't moved all evening; it obviously belonged to Shanker or Miyakara. There was no reason for the other two scientists to touch it.

• • • • •

"We have a dinner guest."

Julian looked up. An animal was standing in full view, about thirty feet away, watching them. It was the same kind as the ones that had killed the infant *Triceratops*. This time he had a much closer view; closer than he wanted, in fact. It was indeed a raptor, apparently the American equivalent of *Velociraptor*. What else could look so graceful, delicate, and deadly at the same time? It seemed to have no fear of them. Not many things would have posed a serious threat to it, small though it was.

The animal stared at them a moment and then opened its mouth and hissed: a sharp whispery sound like air blowing over an open hole. Carl stood up slowly and motioned Dr. Shanker to do the same. Together they inched away from the carcass until they were standing beside Julian.

"We can't let it take our kill," Dr. Shanker whispered, grasping his spear.

"We may have to," Julian said. "Especially if it's part of a pack."

"Move away," said Carl, quietly.

They walked slowly but deliberately another twenty feet back, keeping their eyes on the animal. It did not look at the meat, but watched them instead. Then there was another sharp hiss: but the animal hadn't opened its mouth. Two more raptors were coming into the little clearing, one with its head low, sniffing the ground. All three approached the carcass. Julian wondered how long the animals would tolerate their presence. They may have been puzzled by humans, not knowing where they stood in the strict hierarchy of carnivores around a kill.

The raptors did not stop at the carcass. Instead, they walked around it and kept coming. One stepped directly on the skinned ceratopsian, staining its clawed foot bright red. But they moved slowly and did not seem poised to attack. Julian thought he and his companions were being told to get lost while they had the chance.

But running was out of the question. Instead, they leveled their spears and stepped slowly backward. If it came to fighting, three on three, what would happen? Julian rather thought the raptors would get the upper hand. The animals certainly seemed to understand the difficulties of attacking horned prey; if they could leap onto a *Triceratops* and bring it down, they could surely handle three flimsy spears.

Julian came up against the trunk of a large conifer; his feet were sinking in the thick brown needles beneath it. He looked up with sudden hope but the lower branches were just above reach.

Dr. Shanker set down his spear. "Can you get up there, Whitney? I can give you a boost." He cupped his hands.

Julian leaned his spear against the trunk of the tree. "Why am I always first?" he muttered, fearful of turning his back on those eerie animals. With one foot in Dr. Shanker's cupped palm and his hands against the rough bark, he scrambled up toward the first branch. Still, he could not help glancing over his shoulder.

Two of the raptors were still advancing. The third, hungry or just plain greedy, had turned and gone back to the kill. It was ripping the gut apart with its jaws, one foot placed on the carcass to pin it to the ground. The sight made Julian's teeth chatter; that carcass could just as well be him.

Stretching up with renewed urgency he got one hand around the lowest branch. He was still pulling himself up when the support of Dr. Shanker's hands disappeared from under his foot. He swung free, dangling from the branch, craning his head around to see what had happened. One of the creatures had rushed forward. It leaped off the ground and lashed out at Carl with its hind foot, the huge middle claw extended.

Carl did not lunge with his spear; he simply stepped back, and the animal did the same as if it had intentionally missed. The two stood facing each other, both tense, but neither obviously threatening. That was the warning; the next attack would be for real. Dr. Shanker had snatched up his spear and was standing beside Carl, glowering at the two raptors.

In a panic Julian struggled up further to straddle the branch. Then he hissed at the other two to hand up a spear; he could hold one end while they climbed up the shaft. But neither of them wanted to turn their backs to their attackers. They stood still, tense and ready, spears forward. There was not enough time for both of them to get into the tree.

The silence was interrupted by a loud tearing and breaking of branches. The two menacing raptors backed off, hissing, and then turned and vanished among the gray trunks. Julian saw a huge form moving toward them. With his view obscured by branches he could not see the entire animal but only its lower part, the hind legs and clawed feet. It stooped over the carcass and pinned the third raptor to the ground with one monstrous foot.

Julian felt something grab his arm and he yanked it back in a panic, stifling a cry. But it was only Dr. Shanker's hand. He was standing on Carl's shoulder while trying to find a good place to grip the branch. When he had hauled his bulk into the tree by main strength, he reached down for his spear and then steadied it while Carl clambered up the shaft.

As soon as they were all three on the lowest branch Carl said, "Higher." Julian didn't wait to be told twice: he scrambled up until he was about thirty feet above the ground. Soon they were all perched up there, looking down on the scene below. Even viewed from its own height the size of the creature, so close to their tree, was unbelievable.

"Corla," Julian whispered, hardly daring to breathe. Carl nodded. Dr. Shanker said nothing, but his eyes were wide and staring. He had not been prepared for this close-up encounter.

They watched Corla consume both animals, the live one and the dead one, with all the gore that had been attributed to dinosaurs in popular movies. She tore off huge chunks and swallowed them whole while the animal kicked around, pinned to the ground under her enormous foot. Eventually the raptor stopped quivering and was finally reduced to disconnected parts scattered on the bloody ground. The smell was sickening. Their savior took no notice of them but squatted down to rest after her meal. It looked like they would be in the tree for a while.

"Whitney," said Dr. Shanker, in a hushed whisper, "what happened to your theory of carnivores being relatively scarce, rarely seen? We just encountered four of them at once."

"It's not a theory," Julian whispered back. "It's an ecological principle. But it describes a probability only; anything can happen."

"Very useful 'principles' you ecologists have," Shanker grumbled. "In physics, when. . . ."

But Carl looked at him and said, "Be quiet."

Despite the situation, Julian grinned. Not many people addressed Dr. Shanker so bluntly, and his face showed his surprise. However, he had the sense to obey.

The sun climbed higher. They began to sweat. Irregularities in the branch poked into Julian's legs and he could not arrange himself comfortably. Every time he so much as twitched, Carl warned him with a glance to be still. Carl might have been a branch himself, for all he moved; but Julian began to feel every tiny insect and bit of bark touching him. He longed to make a noise, a sudden movement, anything but cling motionless in the silence; at the same time, his heart jumped at the least little sound he or his companions made.

Eventually Corla stirred, sniffing around on the ground and scraping at something with a hind foot. Julian held his breath and tried to be invisible; they were not much higher than eye level for her, and he did not want to end up as dessert. She bent her head and picked at her teeth with her good forelimb. The gesture would have been endearing in another animal; in fact it was almost ludicrous in this one, because the limb was so small relative to the head.

When she straightened and swung her head about, taking in the surroundings, they caught a sickening blast of carnivore breath. But she either didn't see them or didn't care, and she turned and wandered away. They waited several minutes and then climbed down.

The ground looked like a slaughterhouse. There was nothing left of the meal they'd hoped for. Having no desire to stay in such a dangerous place, they took up their spears and hiked away.

After walking in silence for a while Dr. Shanker mumbled something.

"What?" Julian said.

"God," he repeated, limping along behind Carl.

"God?"

"Yes. You know, that guy up there who looks kind of like me, only his beard is longer. But I think we were taught wrong. I think we just met him, or her, back there."

Julian grinned. Perhaps Dr. Shanker was right: perhaps it was not them, but Corla, who was made in His image.

"But why is Carl helping us?"

Dr. Shanker spoke to Julian across the firelight. The sun was setting. It was three days after the encounter with the raptors and the endless hiking through scrub and rock, gully and hill, had begun to seem like the entirety of life. Carl was away wandering, they did not know where; he had not eaten but had quietly played his strange music and then put the instrument aside and walked off without speaking.

Food had been scarce the last few days but their spirits were revived: in the afternoon they'd found a flat rock on which thick plant stems had obviously been crushed. A small round boulder, the grinding tool, lay beside it. The stringy stems were split and the soft edible insides had been removed. They collected more of the plants from a swampy patch of ground nearby, and made a sparse dinner. Dr. Shanker called them asparagus.

"He could just as well have told us the way," he continued, licking his fingers after eating the last piece of vegetable. "We would have managed on our own. Not that I'm complaining about his presence," he added hastily. "Although it would be nicer to have that friend of his off our tail; she's beginning to give me the creeps, even when there's no sign of her. But why would he want to come all this way with us?"

Julian shook his head. From the start Carl had accepted their travel urge as perfectly natural. He even seemed to know just where they hoped to go, although the caves he referred to were farther north and not far enough west by Dr. Shanker's estimate. It was almost as if he had been stationed on his "sentinel mound" to wait for the time travelers and lead them to the right spot. A romantic notion; and falsified by the fact that they'd already traveled most of the way without a guide.

When Julian tried to ponder the concepts of time and probability, and the mystery of Carl's presence, his mind became confused. As a paleontologist he was more of an observer than a theoretician; so he avoided confusion by accepting Carl's existence at face value, since clearly the man was here and no amount of speculation could argue him away.

Trying to piece together all the clues Carl had dropped about his origin was like fitting together the fragments of bone dug out of a fossil site. It required the same kind of patience.

Dr. Shanker, however, was not so patient. When Carl returned he began renewed questioning in his blunt way, refusing to be put off by the short answers and fragmentary information.

"We're glad of your company, Carl," he began. "But what do you get out of this trip back to your caves?" Carl's response would have come in time, but Dr. Shanker was not one to sit and wait for a slow answer. "Why did you leave your home to show us the way?"

"I am the only one left to show you," came the reply, at last.

"Yes, I'm sure you're the only one left. But perhaps we could find our way alone. We have a pretty good idea of where we're headed."

"Will you know the correct place when you get there?" Carl asked, with his direct, expressionless look.

Dr. Shanker stared back, the twig he was using for a toothpick jutting out of the corner of his mouth. Then he laughed uncomfortably. "Of all the . . . as if he knows where we're going!"

Julian said nothing. Any word from Carl was a potential clue, another tiny fragment of bone sifted from the obscuring dust.

"And where exactly," Dr. Shanker continued, "are you taking us? Is there a particular spot you have in mind? An 'X' drawn on the ground?"

"We are going to the caves."

"Why do you think we want to go there?"

Carl pointed to the moon, faint now, giving only a small amount of light for a few hours after sunset. "Keep count of the moons," he said. "When this one is small, you must be there."

Carefully, as if lifting a tiny, delicate fossil from the earth, Julian put in a question. "Do you go each year at the same time, on this moon?"

Carl shook his head. "I am old, and alone. I no longer go every year. It is where they are buried, under the stones of the hills." In a low voice, so that Julian wasn't quite sure of the words, Carl added, "I would lie under the stones too. But I am the last."

"But this is crazy," Dr. Shanker interrupted, obviously not hearing the last words. "It's just some mumbo-jumbo his grandmother taught him, nothing to do with us. What does he know about temporal reversion? We have only one chance, this year. Forget this every year business."

Dr. Shanker was right, if not strictly polite. It made sense that Carl's forebears tried to revert in the same calculated location, but the time window was a one-time deal: it did not open again each year. If they did not reach the correct area soon they would spend their lives in the Cretaceous. And what of Carl? Would he choose to stay, if offered the possibility of being transported to the twentieth century? Would he even understand the offer? "How far are we from the caves?" Julian asked.

"Two days. Two days until the new moon."

The jagged mountain ahead began to dominate the horizon. Its base was several miles wide with a lopsided, green peak jutting up like a narrow chimney. It was not tall compared to the Rockies in Julian's own time; but against the surrounding low hills it looked impressive and craggy. Carl meant to head for the northern side of its massive base, where his caves were. By Dr. Shanker's calculations, however, they were tending too far north and should be making for the hills south of the mountain.

Well after sunrise, with the mountain shining unbearably in the reflected light, a steep slope suddenly blocked their path. It was a fault scarp with sharp stones jutting out here and there. Carl studied it, looking for a way up.

"Whitney," said Dr. Shanker, taking Julian's arm and pulling him back. "I know you like him—Carl. But it's time we told him where we really need to go. Our time window is very small; maybe as small as a few hours. And our estimate is only good to about four days. We should get ourselves to the center of our estimated diameter of location, not somewhere on the fringes. These caves of his are too far north. I don't think they're in the right place; his ancestors clearly didn't revert. He wouldn't be here if they had. And what about Yorko and Hilda? Did they head more north or south? Or somewhere else entirely?"

"I'd hoped to catch up to her by now," Julian admitted. "But I can't help feeling that Carl knows what he's talking about, and that we should follow him. If he finds the easiest path, it's also the most likely path for Yariko to have followed. Don't you think?" He was really trying to convince himself. "And as for getting to the right place, I don't believe we can do it. We could be fifty miles off either way, and not know it. Plus, you say the lab has to be intact and the program actually running for us to revert. A little more north, a little more south, we can't know which is better, can we? Maybe it won't make any difference."

"Whitney," Dr. Shanker said, frowning while his bushy eyebrows moved up and down, "you're telling me now that you think the journey was useless. We've come a thousand miles and you tell me that—"

"All I mean is that the Boulder Batholith is an underground structure and that. . . ."

Julian felt a tremor under his feet and grabbed the nearest tree, a dead spindly trunk rising up beside him. An earthquake or volcano would send the entire cliff face down on them in a slide of boulders and gravel. But the land looked unchanged in the bright morning sun: the intense green of the bushes, the gray and black of stone, the blue of the sky.

Dr. Shanker's eyes had gone wide. "Damn that thing," he said.

"We don't have anything to feed it this time. Except ourselves."

The vibrations were not the rolling, low frequency shock waves caused by an earthquake. Something enormous was walking nearby, maybe just around a bend in the cliff. Julian looked around wildly for shelter, but they'd left the last stand of trees half a mile behind and only the cliff loomed beside them.

The ground trembled again.

"Can we climb it?" Julian said, peering up at the cliff dubiously. It was covered with loose gravel and looked like a dangerous climb.

"We must, in any case," Carl replied.

They began to scramble up the steep incline. The rocks were hot to the touch in the full sun. The dust and gravel shifted under their hands and a few rocks came loose and rolled down behind, kicking up a wake of dust that hung in the calm air. Nothing heavy could follow them up such a slope; it would tumble down in a landslide. A few stunted, dry bushes twisted out of the gravel, supplying handholds, but the climb was not easy. Julian slipped once and skidded backward, clutching at a bush. The roots tore out of the loose, pebbly ground. When he finally stopped against a rock he was bruised and cut, and badly frightened. He clambered back up to the others and continued upward.

Carl paused on a narrow ledge. Above him the scree continued for another twelve feet, and above that the ground leveled out into more solid rock. But those last twelve feet looked impossible. They were entirely free of bushes or large rocks, and the gravel was so fine that it came apart in their hands.

"Stuck in full view," Julian said, looking around. "We can't even go back and look for a better place to climb. We'll never make it back down." He wondered if Corla's walking could start a small avalanche. They might be buried alive, and she'd never even know it.

"Nonsense," Dr. Shanker said, grinning. "You've obviously never had wilderness training." He leaned his spear against the slope, bracing the point against the ledge, but the end of it only reached halfway up. "Do you trust my grip?" he asked. He knelt on the ledge and lifted the spear, holding it firmly in both hands, until it reached to within a few feet of the top. "Climb onto my shoulders and up to the top. Whitney, I know you hate being first, but you're the lightest."

Julian imagined Dr. Shanker losing his grip, and himself swinging outward and tumbling uncontrollably down the slope. Glancing back down the way they'd come, he was surprised to see how steep the grade was, and how many points of rock knifed out of the gravel.

After an instant's hesitation he took a good hold of the shaft and climbed as quickly as possible. Dr. Shanker grunted; the spear swayed sickeningly. But in a few seconds Julian reached the top and clutched solid rock. He pulled himself up and looked down at the other two.

Carl climbed next, while Julian reached down and steadied the top of the shaft. Then Dr. Shanker handed up the other two spears. "Do you think you can hold my weight?" he said to Carl.

Carl only smiled, faintly. He braced his feet against the slight irregularities of the ground and grasped the spear, dangling it over the edge. As Dr. Shanker put his weight on the bottom end of it, Julian could see the wiry muscles bunching and moving on Carl's back; but he did not lose his footing. Dr. Shanker climbed up the shaft, hand over hand, and was soon standing beside them.

They looked back down the slope to the ground below. Nothing moved; the *T. rex* was not in sight.

"After all, she's used to people," Julian said, mostly to reassure himself. "She's not so different from Hilda, in a way."

Dr. Shanker snorted. "Gonna put her on a leash and bring her back with us?"

"Corla will never be able to climb this, anyway," Julian said. He was panting, but his heart was steady now.

"No," said Carl. "She will find an easier way."

Dr. Shanker scowled. "I'd almost rather she walked up and introduced herself, instead of skulking around. But I suppose she means us no more harm than she did the last time."

"We do not know what she means,' Carl said.

The theory of evolution has frequently been misused to support political and social movements. In the nineteenth century, evolution, for those who believed in it, was thought to represent the struggle by which life gradually ennobled itself. In the early twentieth century, a similar notion was used to support chilling political agendas. Eugenics was based on the notion that a species should "improve" itself by weeding out the "disadvantaged." But the modern theory of evolution, stripped of any progressivism, any concept of a predestined path, is without social, moral, or political application. It is ethically blind, morally vacuous; a description of what is and has been, without recommendation about what should come next.
—*Julian Whitney,* Lectures on Cretaceous Ecology

– TWENTY-TWO –

Carl walked on through the scrub, through Julian's exhaustion and Dr. Shanker's limping soreness; Carl walked on through the scrub, always ahead, fast-pacing, silent.

After the bright morning the sky began to grow overcast. A dim grayish light settled around the bumpy landscape. Somewhere above the clouds the sun reached its zenith, but it remained hidden. The gloom and the damp air added to the sense of stillness, but it was a tense calm, the atmospheric calm that comes just before a rain storm.

Julian had been scanning the hard ground for footprints as he always did, hoping to find some sign of Yariko; and from the corner of his eye he caught a faint, tiny motion in the rocks. It turned out to be a few tufts of fur, sticky with blood, moving in the slight breeze.

"What is it?" Dr. Shanker asked, kneeling beside him.

"The remains of someone's breakfast," Julian said. "It's fresh. The blood is still wet."

They looked about for Hilda's prints; but the predator could have been any small carnivore.

A light drizzle began to speckle the ground and soak into the dust and the porous surfaces of the rocks. Carl moved ahead, sometimes out of sight over a rise in the ground, sometimes visible, striding with his bundles on his back and spear in hand. He looked like a natural part of the landscape; it was easy to forget that he didn't belong in it any more than they did.

The day grew darker as the afternoon wore on. Thunder rumbled in the distance over the western hills. Far away black clouds moved, occasionally lit by faint lightning. Still they plodded on. Carl was confident about his direction even with no sign of the sun to steer by, and he was also sure they were following Yariko. After some hours they took a short rest and sat on the ground with the poncho hides over their heads to keep off the drizzle. The hides gave off a strong musty odor in the damp; the last of the dried meat that they munched tasted like the leather sacks. The sun went down unseen.

"Are we close to the caves?" Julian asked wearily. It seemed that he'd been walking for a lifetime over these gray rocks, with the image of Yariko always in front of him. Sometimes, just before he fell entirely asleep, she would take on mythical proportions: a trail-blazer striding along with spear and knife, defying raptors and perhaps even *T. rex*, showing the rest of them the path.

"Ahead there is a large rock, and then a cleft," Carl said. "We follow the cleft. From there. . . ." He turned his head and looked to the north. "One day's fast walk to the cliffs."

At that moment, they all heard a soft but distinct sound: the sharp clunk of a rock hitting another rock. Dr. Shanker and Carl stood quickly and they all listened; but the sound was not repeated.

"Are there any more steep slopes?" Julian whispered. "Rock slides?"

Carl shook his head. "Only an animal moving would cause that sound."

Julian concentrated on feeling vibrations from the ground, but with the thunder rumbling he could tell nothing. Cautiously, they picked up their spears and bundles and moved ahead, in the direction of the sound. The landscape seemed open, but there were enough ridges and fault lines that even a large animal could have been hidden nearby.

After only a few steps Julian bumped a loose rocks and caused the identical sound, kicking one rock into another. Then two things happened: there was an unmistakable tremble in the ground that was not thunder; and, from well ahead, came a single deep woof, magnified by the bare slopes.

Carl stiffened at the sound, his face intent and immobile. But Dr. Shanker lunged ahead with a shout. "Hilda! Here, girl! Hilda!"

Julian followed him, running uphill over the uneven ground toward a mass of rubble dimly seen ahead, a former rock slide held in place between the arms of two jagged cliffs. As they neared it an object detached itself and came flying down the slope. It was Hilda, now barking madly, nearly turning a somersault as she hit the bottom.

Julian looked wildly around. There was no sign of Yariko. With a sick feeling in his stomach he called her name, but his voice could not carry over the thunder and Hilda's ecstatic barking. Not knowing what else to do, he ran to the pile of rocks and began clambering up. He was nearly frantic; the thought that something might have happened to her in the last few days drove him wild. He called her name again; this time, there might have been a faint answer.

The sound came from just below him. Julian turned, lost his footing as the rocks shifted beneath him, clutched frantically at the air, and slid down the slope.

Someone was calling his name and shaking him. The voice was insistent. He didn't want to move; it was too hard. At last he opened his eyes. Yariko was bending over him and she seemed to be moving, going in and out of focus like the rocks behind her. The ground was spinning and shaking.

Julian wondered with some irritation why she couldn't hold still so that he could look at her. Reaching up he grabbed onto her shoulders and blinked, but couldn't focus. Something warm and wet was creeping down his face.

"Jules, Jules, you're okay," Yariko was saying.

"Keep still," he mumbled, and finally the world steadied, and the ground stopped heaving.

Julian sat up, still clutching at Yariko, staring into her face. He felt blood trickling down his own face but he didn't care. In the days since they'd been separated Yariko had become something of a phantom in his mind: a composite made up of the many images he had of her since their first meeting, and several that came purely from imagination. Now she was there in the reality: in a filthy, tattered shirt, with scraggled hair, puffy eyes, and a dried streak of either mud or blood on her cheek.

Julian felt faint all over again. He was not prepared for the reality of her presence.

Then the rest of the world intruded itself. Hilda was barking and Dr. Shanker was shouting something. A rough hand grabbed Julian's arm and yanked him to his feet.

"Get back," came Carl's voice, sharp and urgent.

Julian looked up and took an involuntary step backward. In the gray light fast turning to black, an enormous shape was moving toward them. Its knobby hide and gray-brown coloration blended into the rocks, so that he perceived it one piece at a time: the curve of the back, a foot, the long tail, and then the head. Once again he heard that vast breathing; only this time, there was no sense of dreamy moonlit peace. One foot was lifted, then slowly set down again, and Julian's own feet registered the vibrations transmitted through the rock. The storm was much closer now, and in a flash of lightning the whole animal became sharp and clear for an instant, appearing to leap out from the rocks.

Yariko clutched his arm and dragged him back. Still dizzy and confused, Julian stumbled against her; but she held him and guided his feet, and they walked backward until they were standing beside Dr. Shanker. He stood holding Hilda's collar and from her throat came a rasping growl, while the hair all along her spine stood on end. Blood trickled from Julian's face.

"Why doesn't he move off?" Yariko's fingers dug into his arm. She did not seem at all surprised at Carl's presence. Maybe, in the intensity of the moment, she had no time to think about how improbable he was.

Julian wanted desperately to call to him but was afraid of provoking Corla into attacking. Carl stood facing the huge beast. His feet were spread wide and firmly planted, and, Julian realized suddenly, he did not have his spear. Corla shifted her weight and lowered her head toward him, stretching her neck; but she did not come any closer. She seemed unsure what to do.

Then Hilda broke free and leaped forward with a snarl. Dr. Shanker cursed, dropped the broken collar, and lunged to grab her, but she evaded him. She rushed past Carl and stood facing the creature, barking and snarling. She was quivering with fury, lips drawn back over her teeth, looking dangerous enough in her own right. The tyrannosaur drew back a step and glared down at the puny thing that made so much noise.

Dr. Shanker crept forward; he was trying to sneak up on Hilda and drag her away. But as he approached, Carl spoke without turning.

"Go," he said. "She is confused now. She will not follow you, and soon it will be dark. Go."

"I'm not leaving him to be eaten," Julian said under his breath, and started to inch forward.

Yariko kept close beside him. She whispered, "He's talking to it."

Maybe Carl was hoping to calm her with his voice and prevent Hilda from angering her. "You should go home," he was saying. "This is not a good place for you. Little food, dangerous cliffs and loose ground. You cannot live here. Go now." He must have spoken to her many times before, Julian thought. The sound of his voice must have been familiar to her.

In the dimness Julian could no longer make out details of motion, but he could feel Corla's uncertainty like a palpable thing. She might lunge forward in attack, provoked by Hilda's threatening attitude, or stand where she was and let them walk away, as she had let Carl do so many times before. He thought he saw Carl stepping backward, but it could have been a trick of the shadows.

Lightning crackled again, and for an instant Julian saw the four figures, motionless in the flash, frozen in action: Carl, Hilda, Dr. Shanker; and Corla, towering above them. Darkness again, and he could see nothing for a moment, blinded by the flash. When vision returned Hilda was dancing around the creature, barking, and Corla's stance seemed to be changing from uncertainty to slow anger.

Dr. Shanker now faced Corla with his spear while Carl, amazingly, stood with his back to her and faced Dr. Shanker. He must have been trying to turn Dr. Shanker away, to prevent him from angering her. But it was too late for that. She rose to her full height, opened her jaws, and screamed. It was an utterly nightmarish sound, high pitched and ending in a loud hiss.

Julian staggered backward and then turned to run, still looking over his shoulder. Over the crash of thunder echoing off the bare hills Dr. Shanker was shouting, but the words could not be heard. Julian's legs were trembling as he ran, and when the screaming hiss came again he nearly fell down in terror. This was no longer the animal they had become used to, a docile, dim-witted beast, hoping to scavenge a meal from the humans. She was *Tyrannosaurus rex*, the monarch of her world, afraid of no living thing.

Then Carl's voice rose above the din: "Stand back!"

The air crackled with electricity.

The sky split open again and revealed Corla moving forward, towering over Carl, her head stretched out above him. Hilda cowered behind Dr. Shanker. The animal screamed again, throwing its head back on its thick neck; and then, just as darkness came down once more, the puny forelimbs came together like a vice around the man beneath her. Julian turned and rushed toward them in a rage, forgetting all about fear. But Dr. Shanker's arm shot out and knocked him off his feet. Stooping, Shanker lifted a large mass in his hands, a rock twice the size of his head. He flung it with a tremendous heave that brought him to his knees.

It might have struck near the shoulder but it was difficult to see. Corla reared upright, the man dangling from her forelimbs, a shapeless bundle, limp and swaying as she moved. Dr. Shanker's next rock, not so large and thrown from several feet closer, struck her on the chin. The great head turned toward him. Julian could not tell if any damage had been done or if she felt any pain at all; then in the next flash of lightning, now dim and far off behind the clouds, he saw blood gleaming where the rock had struck.

He felt more than saw Yariko beside him hefting a small rock. The two threw their rocks together. Julian's was too low, and landed on one great hind foot. Yariko's curved up and hit with an audible smack below the right eye.

"It's going to eat him!" Yariko cried. Julian tore the leather sack off his back; it was empty but for crumbs of dried meat, but he threw it anyway. It was not heavy enough; it landed somewhere on the ground, unseen in the darkness.

This time Corla responded. She dropped the limp bundle, which landed on the ground with a sickening thud. She lunged toward Yariko and Julian. Her first step covered half the distance. They turned together and ran, terror overcoming any thought of fighting. Julian had to force himself to stop and turn, with a wild thought of trying to delay the animal; he might save Yariko's life at the cost of his own.

But Dr. Shanker was there first. He came in from the side, running, and stabbed his spear deep into the animal's hind leg.

It was not a terrible wound; it was the only place the man could reach without getting beneath her, under the arch of her belly. But she felt it. She lowered her head and swung it with horrible speed, striking Shanker and lifting him from his feet. His body arced through the air and hit the ground several yards away, the force of the impact sending him rolling, limbs flailing like rags, until he came up against the foot of the rock slide. The animal continued to turn her head, hissing now and biting at the spear, which looked like a tiny sliver jutting from her massive thigh.

And then, inexplicably, she began to walk away. Yariko had just reached Dr. Shanker, but seeing an opportunity she launched herself on Hilda and dragged her down, clinging to her hind legs so she wouldn't follow Corla and provoke her again.

The huge shape, now only a deeper blackness against the gray-black sky, paused when it reached the spot where Carl had stood and faced her. She bent her head to the ground as if sniffing something. Then she stepped backward, gave a rumbling growl that sounded like a landslide, and, turning, disappeared into the night.

Julian ran forward to find Carl, caught his foot on something softer than rock, and sprawled on the ground. The stones were wet and sticky.

"Where is he?" said a voice behind him in the darkness.

It was a harsh voice, distorted, barely recognizable as Dr. Shanker's. Julian stood and reached out, feeling in the dark, and clutched at the man's arm. He meant to help him, but instead found himself leaning on Dr. Shanker.

He was shoved aside, roughly. Dr. Shanker knelt beside the body.

"Here. I can't tell with all this damned blood in the way. Put your hand here. Do you feel any pulse?"

Yariko knelt too and let her hand be guided.

"Yes," she whispered after a tense silence, and hope came back to Julian. He knelt down beside them.

"Carl?" he whispered. But there was no answer.

We humans have developed a comforting, self-centered philosophy that helps to shield us from the dreadful truth. We put a high value on individual life, liberty, and happiness. But in the broad context of the history of the world, life is cheap and death is the only certainty. All species become extinct. Habitats change. Ecosystems develop and then collapse. Our own civilization will be gone soon enough, the way of Triceratops horridus, *the way of* Tyrannosaurus rex.
—*Julian Whitney,* Lectures on Cretaceous Ecology

– TWENTY-THREE –

They did not see Corla again. Whether she returned to the river valley to live out the rest of her solitary years, or died of thirst and hunger in the lifeless hills, they would never know.

They did what they could for Carl, groping in the blackness, unable to light a fire in the open rain. There was not much hope. That grip must have crushed his rib cage; there were likely terrible internal injuries. It was incredible that he was alive at all.

Dr. Shanker also needed attention. He pushed the others away and savagely insisted that he was fine, while refusing to leave Carl's side; but Yariko and Julian managed to sit him down and feel for broken bones. Amazingly, there were no more than one broken rib and a gouged shoulder where he'd hit the ground; that and a multitude of minor bruises and cuts. He coughed frequently. The morning would show a number of teeth knocked out and the side of his face badly scored. But for the remainder of that long night they could only listen to his hoarse breathing and distorted speech.

The three of them huddled on either side of Carl and covered him with skins, knowing that he needed warmth during the cool part of the night. The storm receded to a low murmur of thunder now and then. Yariko sat across from Julian with her hand on Carl's. She still had not asked who he was.

As the grayness of dawn came at last, he woke. Yariko gave him some water, and Julian told him they would carry him to the caves and take care of him until he was well.

He smiled faintly. "You will get there in time," he murmured. Julian had to lean close to catch the words. Bloody froth stood in the corners of Carl's mouth and his breathing was labored.

A thin drizzle fell. Carl's face in the cold light was ashy and drawn. Yariko held his hands to warm them. He looked up at her for a long time, as if studying her face, and then he closed his eyes again.

Julian had never watched this kind of death before; but he knew the instant Carl's spirit, his consciousness, was gone.

The whole world seemed hushed as the sun rose in a thin strip of clear sky below the clouds. They sat on and on, not wanting to believe. Even Dr. Shanker was silent; even Hilda lay still. The sun climbed until it shone out, lighting the underside of the clouds above. No bird sang to break the silence of a Cretaceous dawn.

Yariko slowly released Carl's hand. Dr. Shanker spoke first. "I'll gather some rocks. We'll build a cairn." He rose painfully, spat out a wad of thick blood, and called to Hilda.

"No," Julian said.

Shanker turned in surprise. "No? You want to . . . leave him here?"

"We're taking him with us."

"What? Whitney, that's. . . ." He lowered his voice. "Let's be reasonable, now. We can't carry him that far. There's no conceivable purpose—"

"We're taking him to his caves. He wanted to be buried there, with the rest of his people."

Julian rose and collected the four spears. One was coated in dried blood, Corla's blood. She had pulled it out of the wound with her teeth. Using the ponchos and several leather strips from Carl's sack, he fashioned a litter and laid it on the ground beside Carl. His two companions watched him in silence.

"This is just sentimentalism, dangerously impractical," Dr. Shanker finally said in his new, slurred voice.

But Julian was beyond reason. When he didn't answer Dr. Shanker said angrily, "You know you won't get far with all that weight."

Yariko stepped forward to help. Working gently, they slid Carl onto the litter. Yariko straightened his arms and closed his eyes, and then covered the body with skins. Julian crouched down and took one end of the litter and Yariko took the other. But Dr. Shanker roughly pushed Julian aside, muttering, "You and Yariko share the back end. It'll take two of you." Then lifting his end, he said, "If we're going to do this insane thing, let's get going."

It took all that day and night to reach the caves, following the descriptions that Carl had given. Julian set his mind on the task and let no other thought or feeling intrude; he could not have gone on, otherwise. The litter was heavy, and Yariko and he together could barely manage what Dr. Shanker carried alone on that rough ground. They stopped to rest often, just long enough to catch a breath and mop the sweat out of their eyes. At each stop Dr. Shanker looked more bent and savage. They told him to leave it; they told him to take a rest, and let them do the work; but he ignored it all. And although he was slow, sometimes staggering from the pain, Julian knew they could not make it without him.

Julian saw nothing of the landscape around him. He felt the sharp ground cutting into his feet through worn-out soles. He felt the heat, and the painful weight of the poles pressing on his aching shoulders; but his mind was numb. The day passed in a blur and ended in blackness. The moon came up, a thin crescent above the western horizon, so dim that it hardly outshone Venus glimmering beneath it. By this faint light, they kept to the path.

At a steep downward slope Yariko wanted to pause and re-position but Dr. Shanker ignored her. He was breathing in gasps and spitting blood every few steps; they were all exhausted.

"Stop," Yariko begged. "Not for me, for you. You need a rest. You won't make it."

"If I stop now I'll never get up," Dr. Shanker growled. "Later— not here." He started down the slope at a stumbling shuffle.

The ground was rough and Julian staggered as he walked, although he was on the upslope end while Dr. Shanker remained in front taking most of the weight.

Near the bottom, Shanker stumbled and almost fell. The litter tipped but with a great effort he kept his balance, straightened his end, and made it to the bottom. Once there, he set the litter down and collapsed without a sound, clutching at his chest. Hilda licked at his face and whined.

Julian gave him the last few mouthfuls of water from the skins. Then they opened one of the sacks and wrapped it tightly around his chest, hoping the pressure would ease the pain in his ribs. He allowed himself only a few minutes to rest, and then insisted on taking up the litter again. Yariko silently moved to his end and took one side from him. Julian staggered on at the other end.

Just as the sky was brightening, they came to the base of the little mountain. It did not have a gradual slope, but rose suddenly out of the rocky rubble, nearly vertically, cut by fissures and tufted with bushes. In the diffuse predawn light the cliff looked immense, rising into the sky, but it was in reality no more than a hundred feet tall. Here and there a small tree grew sideways out of the rock.

Coming closer, Julian saw that the wind and rain had eroded the cliff face into shallow caves. Two looked quite large, extending well back into the cliff. A little to one side, eerily sending out long shadows in the dawn light, were the cairns of Carl's ancestors. There were twelve in all.

They had found Carl's caves; they had brought him home. But it was not time to rest. As the sun rose they worked slowly, tiredly, to scrape at the hard earth and collect loose rocks. Carl was wrapped in his poncho and laid in a shallow depression under a high cairn, in the midst of the others.

Yariko had tears on her cheeks as she placed the stones, little by little covering the body, until only glimpses of leather showed, and then nothing. Julian wondered at this, since she had not known Carl, but he didn't question it. He felt as if the whole Cretaceous world was grieving for their fellow being. For himself, it was too hard a task. He could not lay the weight of stones one by one on his friend, his mysterious teacher and guide in the Cretaceous wilderness. Even when the covering was complete he could only bring new rocks over and place them on the ground for Yariko and Dr. Shanker to lift onto the growing cairn.

A few times over the past week Julian had been sure they would all be killed, bungling people from a future age, and Carl would be left to return to his gentle animals and his solitary life. It had never entered his imagination that Carl would be the one to die. Even after the attack, although he knew no human body could survive that much trauma, Julian still half believed that Carl would get up, say they must go now, and stride ahead with his spear. Anyone who could stand and face *Tyrannosaurus rex* with so much inner strength must surely be able to face off death itself.

But in the end Carl was as fragile as any of them. His enclosure, his hill, his ingenious water system, the fire pit and the marvelous spices hanging from the ceiling, were all without their master now. The animals would die of starvation, milling about the enclosure, waiting for their caretaker to return. At that thought the tears began to trickle down Julian's face, dropping on the ground, evaporating quickly on the warm rock. He turned away.

With the last stone in place the energy that had kept them all going for the past day and night was gone. Dr. Shanker could barely stand. They sat down just outside the entrance of the nearest cave, shaded from the intense morning light, and leaned back against the stone cliff. Hilda lay down and put her head in Dr. Shanker's lap. For a long time they sat without speaking, too numb even to doze.

"Now that we're here," Dr. Shanker said at last, "where are we?"

His question was to the point; and two days ago, it would have mattered to Julian. They could be miles from the calculated spatial window. Their path could have been thrown off considerably by the gullies and other obstacles that had been circumvented. "Here," however, was as good to Julian now as anywhere else.

He thought back to their second day on Cypress Island and remembered seeing a new crescent moon through the trees. Again, on the river journey, he had seen a new moon. And in the predawn twilight, as Carl slipped away, a thin moon was rising again. It was exactly two months since their entry into the Cretaceous world. Carl had been right.

And the four of them were relatively unscathed after their thousand-mile trek. Dr. Shanker was still in pain, but his breathing seemed easier and he had stopped coughing up blood. His ankle was nearly healed although, as he said, the toe of his dilapidated sneaker would bear teeth marks to the end of its days. Yariko had a long scrape on one arm and a gash on her cheek, and the palm that had been bitten by the dromaeosaur, so long ago, had a wide pink scar that didn't like to be stretched. Hilda's face where she'd been kicked would always have a little bald spot, and she didn't seem to see as well through that eye. As for Julian, he had some minor cuts and an aching head from hitting a rock when he'd fallen down the gravelly slope. But they had gotten off lightly, all four of them, considering their adversary. Now there was nothing to do but wait.

"Yariko," Julian said, turning his head to look at her, and taking her hand. "Tell me. Tell me what happened when you were alone."

Yariko looked far from ready to begin her story. "There's not much to tell," she said in a tired voice. "And some of it I still haven't figured out completely . . . well, never mind. I don't want to think about being alone."

"Tell me some of it," Julian urged. "You can finish later. I want to know."

"OK. But you've probably heard half of it already, from Dr. Shanker. How we lost you in the river."

Julian nodded.

"And then that Triceratops chased us up stream, and we couldn't come back to you for a long while. By the time we did, it was too late. You were gone. We searched everywhere. It was very strange. When we finally left the river we thought we were following your trail, and we were happy enough to find the pile of wood you left behind in a little woods. We used it that night. We were impressed, you know. Not only did you remember to leave us a sign, but you were thoughtful enough to leave something useful."

"You overrated my cleverness," Julian put in. "It was Carl's wood."

"Of course," she said. "It never occurred to me how unlikely it was, you taking half a day to chop wood, stacking it up under a tree, blowing the dust off your hands, nodding your head over your handiwork, and then pushing on toward the west and not bothering to wait up for us. But we weren't thinking clearly. We had other things to worry about. The storm hit us that night, and I thought we were going to be blown and washed into the river and drowned. That's when Dr. Shanker lost himself."

"Quite the contrary," Shanker said, calmly. "It was you who lost yourself."

"No," she said. "I knew exactly where I was."

"As did I," he said.

"As I was saying," she continued, winking at Julian, "I couldn't find Dr. Shanker anywhere that night, or in the morning. I had no choice, really, but to continue toward the west. I was hoping to catch up to you. Hilda and I found what looked like an old river-bed running in the right direction, and it made for better footing.

It was smoother and less overgrown with thorns than the rest of the landscape.

"I'd only been on it for a short time when I found an old leather sack tangled into some bushes. At first I thought it was a dead animal and I was going to scavenge a meal. To tell the truth, I was pretty upset at you when I saw it was an empty sack. I thought you could have at least put something in it for me." Yariko laughed weakly.

"My apologies," Julian said.

"Accepted," she said, squeezing his hand. "After I calmed down and stopped raging at you, I took a good look at that sack and noticed how well it was made. It was carefully stitched together and had a braided leather drawstring. You have marvelous skills, Julian, but I suspect that you're incapable of stitching one thing onto another thing; and besides, you wouldn't have had time. It was very confusing. I didn't know what to think. Later I found another small pile of wood, which I used to keep warm that night. A few days later, I found this."

She reached to her belt, hidden under the ragged hem of her T-shirt, and pulled out a short bone knife. It was nicely honed down to a sharp edge.

"It's a tooth," he said, taking it and turning it over in his hand. "Tyrannosaurus, I think." It was about six inches long, curved and tapered, the edges naturally serrated. A hole had been drilled through the blunt end and a leather thong tied on. "It's beautifully made."

"And very useful," Yariko said as he handed it back. "Quite sharp. But I was perplexed. Either you had been clever enough to make all these implements, or I had discovered human habitation of the Late Cretaceous. I thought the second option slightly less improbable. I'm a scientist, after all, and I had to accept the least preposterous of the two."

"Thank you," Julian said. "It has nothing to do with cleverness. I already have a knife, and for your information I did make my own sack, out of Triceratops hide."

Yariko laughed again. "I'm so proud. Not many people could say that, you know." Dr. Shanker snorted, and she continued. "You should have seen me, Julian. Tattered, hungry, alone in the Cretaceous wilderness, sitting on a rock pondering physics equations.

Hilda thought there was something wrong with me. She kept whining and pawing at my legs."

"What equations? Did you figure out how other people had gotten here?"

"Well, sort of. Not really. But that's all too complicated for now, and I'm so muddled. Anyhow, I had to stop thinking about physics pretty quickly. I had to find food and dodge that horrible animal that was stalking me. I couldn't tell if it wanted me for a meal, or was only curious about Hilda."

"What animal?" Julian asked, but Yariko was speaking again.

"Being alone, maybe—maybe the only human being in the world, not knowing if you were even alive, or if we'd find each other—Julian, I can't describe it. I just can't describe it."

"You don't have to," he said, pulling her close and holding her in a tight embrace. "I know what it feels like. We'll never get separated again. No matter what happens."

Dr. Shanker was already snoring, tipped over on one side. They lay down in the shade, on the rough ground near the cave, and slept.

Science, after all, is nothing more than a rigorous application of common sense.
—C. Shanker

– TWENTY-FOUR –

2 September
5:30 AM Local Time

Mark Reng's eyes were red and swollen, his face gray and bristly. He looked less like a skinny kid now and more like a contemporary. Earles didn't think about what she must look like by now; it wasn't a concern.

She had more important things to focus on. Mark was giving her a nod from the portal.

Bowman was sound asleep, drooling onto the page of an open notebook that served as a pillow. Ridzgy turned from the computer screen with a triumphant look.

"I'm ready to set up the experimental run, if Mark is," she said. "The program's up and working, the parameters are set, and the other machines are standing by to record." She indicated the four slave machines ranged along the bench, each monitor showing scrolling sequences of numbers and symbols, never ending, endlessly repeating.

Please God I never have to watch a program sequence again, Earles thought to herself. "Very well," she said. "You square things with Mark. I'm going to step out for a few moments while you set up."

Ridzgy's composure was admirable. She didn't even blink as she turned back to the monitor.

Earles went out into the dim hallway. Throughout the night she'd encountered several overworked, chronically sleep-deprived graduate students near the coke machine in the lobby; but at five-thirty in the morning the place seemed to be empty. A time of truce, Mark had called it, when graduate students could leave without fear of notice: the hour before dawn, after the equally overworked junior faculty had left and before one's advisor arrived again for the day.

By seven o'clock the junior faculty would be back, the red-eyed graduate students would be back, and the building would come alive again. By nine o'clock, if not before, government regulators from OSHA, Occupational Safety and Health Administration, would be descending on the place to permanently shut it down, having heard through mysterious channels about the "accident" in a lab run under government funding. They would confiscate the evidence and the equipment, take away the funding, and probably end the police investigation.

Looking at her watch, Earles saw that it would be a close thing. She paced the lobby on the first floor, knowing that patience was still needed: Ridzgy must be given time to finalize her own plans.

After ten very long minutes she descended once again to the basement, and the too-familiar lab.

Bowman was awake, wrinkled and grumbling; there were lines on his cheek where it had lain against the notebook. The notebook itself was no longer there. None of them were there. The counter was remarkably neat.

"I just straightened up a little," Ridzgy said with a mirthless smile, following Earles' gaze. "Stacked 'em all in a safe corner. No sense risking all those valuable records, if this thing blows up again."

Earles glanced toward Mark, who stood outside the vault yawning. Their eyes met. Reaching her hand around to the radio on her belt, she pushed the emergency signal button.

"Well then," she said, taking a chair and casually stretching out her legs. "I think it's time to show what we know."

• • • • •

Julian woke to find the others still asleep, except for Hilda who was nosing around and pawing at the rocks. She may have been chasing lizards; but if so, they were too quick for her in the heat. Now and then she sat down and scratched at an itch with her hind foot. Watching her eased some of Julian's heartache, because she was so busy and so innocent at the same time. For her, life went on.

After a while he got up quietly, trying not to wake the others, and began to explore the caves. The one at his back was empty, except for a stack of wood and a few bits of leather, tattered and decayed. It was very small. There were, of course, no bats; they had not yet evolved.

The second cave opened on the cliff face only a few yards away. The entrance was no more than a crack in the rock, wide enough to fit one person. Dr. Shanker might have to turn sideways and squeeze, but Julian slipped in easily. The narrow mouth opened up to a large room, fairly dry and quite safe from big animals.

Whatever Carl may have thought, Julian did not believe that the cave was dug by humans. It had the irregular shape and uneven roof of a natural formation. Somebody had smoothed the floor, however, and carved rough shelves along one wall. As his eyes adjusted to the dimness he saw that there were objects on the shelves.

There was a large knife that was carved from a sliver of bone. A smaller version lay near it. They were not particularly sharp, and were crudely made. Beside the knives lay two clay bowls. There was no clay in these hills. The bowls had come from near the river, maybe brought here by Carl. Beside the bowls was a flat clay pan with a stick for a handle, but it was cracked in half, and useless.

The last object was a tiny leather sack tied closed. Julian turned it in his hand, puzzled, and then began to work on the knot. The leather string fell apart in his fingers. He turned the sack upside down and shook it. A small object fell out into his palm.

There is a fear that comes when one sees something familiar in a place where it is not expected, where, in fact, it cannot exist. This fear came to Julian. His free hand went up against the rough wall for balance. The cave seemed to close in around him.

A shadow fell across the entrance. "What did you find?" It was Yariko's voice.

Julian could not speak. He looked up and held out his open palm.

"What is it?" Dr. Shanker asked, coming up beside Yariko.

"My pocket compass."

"A compass? Whitney, are you telling me you've had a compass on you all this time? Do you realize. . . ?"

Julian shook his head. "I didn't bring it. It was broken, on Cypress Island. Remember? We couldn't find the magnet. Without the magnet it was useless. . . ."

They all stared at the compass, Julian stooping against the rough stone, the other two standing over him. The letters "JW" were scratched into the tiny glass face.

"Must be yours, all right," Dr. Shanker said.

"But how did my compass get here? I didn't bring it. It's still on Cypress Island, somewhere in the trees near the beach, in a million pieces." He thought back to that first night in their new world and the fear they'd all felt; and his own hopelessness, when the compass was broken. Now both he and it were here, a thousand miles from where they'd started; but he hadn't brought it with him.

"You did bring it," Dr. Shanker said. Julian shook his head again, but Shanker continued. "In another time, it didn't break. And you brought it here."

"I don't understand."

Dr. Shanker looked at Yariko, and she nodded. "Yes," she said. "I had already come to that conclusion."

Julian stared from one to the other of them, waiting for an explanation.

"Hold on a minute," Dr. Shanker said. "I want to take a close look at this cave." He squeezed himself out through the entrance.

Julian looked at Yariko. "What do you see?" he asked.

Yariko was staring at the opposite wall. It was quite dim, but now Julian could make out faint tracings on the wall. He walked over and touched them with his fingers. The lines seemed to be painted on rather than chiseled, and they were regular, in even rows extending from waist to shoulder height. He put his face close to the rock but still couldn't see anything clearly.

Then the cave seemed to light up and the lines leapt out at him, startlingly clear and close. He jumped back.

Dr. Shanker had entered with a makeshift torch; he was now standing beside Yariko. "Would you look at that," he said, and there was awe in his voice.

The lines were curved designs separating rows of writing. Whatever the paint material was it reflected the firelight very well; the letters seemed almost to glow themselves. In places the writing was streaked with moisture and with a black mold, but much of it was clear.

They began to read.

The laconic record of a tiny community slowly revealed itself. There were no sentences as such, only phrases and numbers, and some didn't make sense to Julian. The top line he thought he understood.

"Arrival at second new moon since entry. Reversion point unknown. Day 1 after new moon. Day 2. Day 3 after new moon reversion point and time established as W. vanished when in plain sight. Correct place set down although too late. Future retrieval possible? Day 4 will wait here some time in case of retrieval attempt."

The line ended there; below a set of designs, very beautiful designs, the writing looked quite different. It seemed to record three generations of births and deaths, and although Julian could not immediately make out dates or spacing of these events, he thought there were more than a dozen people, perhaps as many as twenty, alive at one time. Apparently not all were buried outside the caves, as there were only twelve cairns. Thirteen now, he corrected himself. Some must have died while journeying, or out hunting, and never been found.

The third line down seemed to record a migration to the east. The fourth and last line, which he had to stoop to read, looked like a record of journeys back to the caves for burials and other reasons that were not clear.

Julian stood, feeling the ache in his legs and back. The words glowing in the light of Dr. Shanker's torch did not solve the mystery. He wondered who "W" was; the lucky one who was in the right place to revert when the time came, apparently; but who? And who were the people left behind?

"Look there," Yariko cried, pointing to another spot, very low on the wall at the back of the cave. "It's a diagram."

They moved closer with the torch.

The diagram was of geometric shapes, and there were symbols, mathematical symbols, around it. There were no words.

"It's the caves," Dr. Shanker said, pointing to one part of the picture. "See, here's this one, and here's the little one. And here are the cairns."

"What's that triangle?" Julian asked, touching the clean shape that seemed to tower over the crudely drawn cairns. "There's nothing like that here. Look, there's a little moon over it."

"It's the reversion point," Yariko said. "See how it's in the center of a circle? I bet that's a roman numeral III next to the moon: the third day after the new moon. And here," she dropped to her knees and bent forward excitedly to look, "here are the measurements to find the circle. It's quite large, actually." She looked up at the others. "We could mark the nearest edge, and this corner, and make sure to be inside the line."

"Yes, but when? Is there a time, or do we sit on the ground all day, waiting?" Dr. Shanker frowned at the rest of the picture. "This one makes no sense."

There was something naggingly familiar about that second picture that Julian couldn't quite place. There was a half circle with lines coming out of its center, an overlapping ellipse, and a few other strange symbols.

"Somehow it's familiar," he said. "But I can't make it out. Maybe later when I'm less tired." He turned back to the wall of writing. Some of it might be a century old; yet the paint was unfaded.

The torch went out and the writing disappeared. Only the designs could be seen, faintly.

"But who were they?" Julian asked, although he knew his companions couldn't answer that question any better than he could.

"Jules, come outside," Yariko said, holding out her hand.

He let himself be led into the late afternoon sunlight. The shadow of the cliff stretched out toward the cairns. They sat down against the cliff face.

"Don't you want to hear about my mysterious physics equations?" Yariko asked.

"But the compass . . . and the writing. . . ." Julian felt as one does after a night of heavy dreams, trying to shake off their influence in the gray of morning.

Yariko laughed and squeezed his hand. "Do you know, you have such a funny expression when you're confused."

He started to mutter, but she cut him off.

"We left off when I was sitting on a rock, puzzling out equations and the existence of other humans in the Cretaceous."

Julian nodded, still skeptical, but willing to hear her out.

"It took me several hours of hard thought to see the possibilities in the third and fourth terms of the Taylor expansion. Beyond the fourth term, I thought the numbers were negligible."

Julian had no idea what she meant but Dr. Shanker seemed to approve. He nodded.

"But even with the equations straight in my head," Yariko went on, "I still didn't believe it. Quite impossible. And not exactly consistent with the facts.

"For several days I thought a good deal more about survival than about physics, I can tell you. Then one night while Hilda and I were curled up together in a bush, I lay there thinking about you and remembering all our little conversations together—I missed you, you know- -and suddenly, like a rock falling on my head, or like that apple of Newton's, I understood. I saw the answer."

Yariko paused and looked up. "Julian, do you remember our first or second night on Hell Creek, lying on the platform in the tree? We made up such ridiculous stories. Do you remember what we talked about?"

"Yes. Of course I do. I thought of the same thing when I met Carl. There was a population of people who survived, at least for a time. Now we know they came to these caves. I've been thinking quite a lot about it, in fact, even if I don't know any Taylor equations. Obviously, someone tried to rescue us, but didn't hit the exact time."

"And they found the bits of your compass in the woods, repaired the thing, and brought it here?" Dr. Shanker snorted derisively. "Come on, Whitney. You're not as slow as that."

"That's true. It doesn't explain my compass. But the writing— they were obviously here, and they didn't all revert."

"There was no rescue party, Julian," Yariko said. "That was never a possibility, and the idea of "future retrieval" by someone else is hardly possible either. The finest of calibrations can pinpoint a location, or a time, with only so much precision."

"Precision meaning how close each result is to the others," Dr. Shanker explained in a condescending voice, as if talking to a child. "A very different thing from accuracy, which is—"

"I know what accuracy is!" Julian snapped. He didn't need a lecture on scientific terms. He was trying hard to follow what Yariko was saying. "Go on."

With a warning look at Dr. Shanker, Yariko continued. "Like he said, each run, even identical runs with identical settings, will give a slightly different result. Another group getting in the vault and arriving within fifty or a hundred years of us, and in the exact same location, would be, well, beyond probability."

"But you were getting samples from the same place, over and over. Those beetles and stones and such. Are you saying they were from all different times in Earth's history?"

"They were all from roughly the same time and place. A few centuries here or there, a few hundred kilometers here or there; that's as precise as we could be. Plenty good enough for translocating similar samples, but not good enough for a rescue team to get anywhere near us."

Julian didn't like the direction his thoughts were going. "Does that mean we can't revert? That the instruments won't be set to exactly the right place?"

"Not at all," Dr. Shanker said. "If we were true Cretaceous objects, we couldn't count on being brought into the vault on a given run, because of the variability. But reversion is different from retrieval, or simple translocation. We originated in the vault, and we'll revert to the vault, as long as the settings are correct."

"So what does any of that have to do with Carl?" Julian asked, bringing the conversation back to where it had started, and where his mind was still focused.

"It means," Yariko said, in a quiet voice, "That nobody but ourselves has come to this place and this time period."

Julian sat very still as he took in her words. Nobody but themselves . . . nobody but myself, and Yariko, and. . . . When he realized what that actually meant, the sense of awe kept him silent for a long moment.

At last he said, "Then Carl would be our descendent."

Yariko nodded. She took his hand and held it very tight.

"The wonder of it," Dr. Shanker broke in. He'd obviously been waiting impatiently for understanding to dawn on Julian, and was too eager to allow time for sentimental feelings. "The initial reaction in the vault sprayed material through the space-time manifold. We were not sent back to a single point in time. We now know that there were at least two Julian Whitneys, two of each of us, that appeared at distinct and separate points in time. We, the three of us sitting here now, have just lived through one set of events; our counterparts, waking up in a slightly more distant past, must have faced a different sequence of events. In one life, one time, your compass breaks; in another, it survives and helps us find our way.

"We'll never know how many other versions of us might have appeared at different times. Probably no more than four, if Yorko is right about the equations. But then what happened to the other two? Mauled and eaten by dinosaurs? The probabilities would tend in that direction. Another version of Frank might have lived; another version of me might have died. Perhaps I did see a Cairn that day on Cypress Island.

"The whole thing is fantastic, isn't it? And yet, never forget the underlying philosophy of science. Observe, observe, observe; then draw your inferences. Science, after all, is nothing more than a rigorous application of common sense."

Dr. Shanker grinned and stuck out his beard in the way he had when making a speech. "Thus. Observation: another human being lives in the Cretaceous. He suggests he was born here and came from these caves. Observation: your compass is here, in the cave, in perfect working condition. Observation: sixty days ago, you left your compass on the sand, a thousand miles away, broken beyond repair. Inference: we have been here before, duplicates of us if you will; or maybe an original set, and we are the duplicates. And Carl is descended from us. Of course there are many more observations to support that inference, but you get the idea."

Julian was silent for a while, trying to take in these fantastic ideas. The sun was low and the shadow of the cliff now stretched over the cairns. His head was whirling.

Yariko touched his knee. "It gives you a strange feeling, doesn't it?" she said. "Jules, what generation do you think he is? He couldn't be. . . ."

My son? Julian thought. My grandson? His chest began to ache. Carl was dead now. And yet, what was time? He himself, Julian Whitney, was dead, and had not yet been born; his birth was still sixty-five million years in the future, and yet he might be lying under a cairn on Cypress Island.

He had a sudden sharp image of Carl leaning back against the stone wall of his hut, a twig in his mouth, talking in his short phrases about the land to the west. Carl had shown them the way.

"He must have been the last of the generations," Julian said, finally. "It would have taken a lifetime to build his hill, and more than a few people. I think he was third or even fourth generation. The writing didn't show years, only generations." He looked at Yariko. "I find it amazing that against all the odds we brought a child, or more than one, into the world."

"A little community, surviving in all of this wasteland," Dr. Shanker said, waving his hand, vaguely taking in the entire Cretaceous world. "Passing on stories from one generation to the next. The seaway to the east. Reaching the caves on a certain moon in this season. But why did they all die out?"

Julian played with a pebble by his foot, not wanting to look up at the cairns now under the afternoon shadow of the cliff. "I think I can guess," he said. "It's unlikely enough that one child survived. A whole group? Think of the mortality rate. Predation. Malnutrition. Not to mention how inbred they would be—"

"He did know where we had to be, didn't he?" Dr. Shanker said, obviously following his own train of thought. "But I'm still not clear on how he—or we—picked this spot. Obviously it's the right place, because someone reverted . . . someone named W," he ended, turning to Julian. "You reverted."

Julian's emotions, so keyed up as he tried to believe that Carl was his great-great-grandson, did a backflip. And if he reverted without Yariko, then—

"No," he said. "I wouldn't leave without Yariko." He looked at Dr. Shanker. "And you couldn't be the father."

"Frank!" Yariko cried suddenly. "Frank Walden. Frank was W."

"That's right!" Julian hadn't actually known Frank's last name, but that didn't matter. "Frank reverted, and I stayed with Yariko. I couldn't have reverted or I wouldn't have been able to teach paleontology to the next generation. Carl knew some paleontology. I'm serious."

"OK, you win," Dr. Shanker said. "Frank would never pass down paleontological knowledge to his offspring. But I still have to be in the gene pool somewhere."

"How's that?" Yariko asked, smiling at his smug expression.

"I may be a physicist, but I remember my basic genetics. Carl had blue eyes. Yours are black, Whitney's and mine are brown. But there are blue eyes in my family. It seems to me that you two together could hardly bring a blue-eyed descendant into the world."

"I could be carrying genes for blue eyes too," Julian said.

Yariko shook her head. "I'm not. No Europeans in my stock. I'm as pure as they get."

"So you see I would have to be in there," Dr. Shanker said again.

"What about Frank?" Yariko put in quietly. "He had blue eyes, and he was young, too. Maybe he . . . maybe before he reverted. . . ." She stopped and looked quickly at Julian. "No, never mind that. He reverted the third day after they got here, according to that writing."

"Or Whitney did; we can't know, can we?" Dr. Shaker said. "What if—"

"What if, what if a lot of things," Yariko interrupted, feeling Julian stiffen. "This is useless speculation, and has nothing to do with us now." She paused. "The question is how long will it be before our window of reversion opens, and where do we need to be when it does?"

"All right, all right," Dr. Shanker said with sudden loudness. Hilda looked up at him questioningly. "We clearly need to hold another council of war. Dr. Miyakara. Dr. Whitney. At our last council, as you recall, under the palm trees of Cypress Island, we decided to try the chance at reversion; and to make our way a thousand miles to the west. How long has it been? Sixty . . . no, sixty-one days? The river was lucky for us. But have we made it in time? Whitney. What's your expert opinion?"

"I don't know." Julian felt suddenly weary at the thought of all those miles, through swamp, river, and forest, over stony hills and across gullies. He didn't want to struggle onward anymore. He wanted to stop and rest, here at the caves, reversion or not. "I guess this would be day one of the second new moon. That would make the reversion time two days from now."

"Very good. Yorko has already told us her opinion. Let me tell you mine." Dr. Shanker stopped to think, scratching at his beard. "A thousand miles is a long way to travel with any kind of accuracy. But we've made the journey before; sixty years ago, maybe? A hundred years? And at that time, apparently, we did have a compass. We came here. We came to these caves and we stopped, and one of us reverted. I put my trust in my previous self: I vote we stay here. We'll either revert within the next few days, or we'll live here forever."

Julian was already losing interest in the discussion. He was hungry, tired, and his head ached. It was getting late; there would be time tomorrow to sort out details. Looking around at the other two in the fading light of the evening, he noticed suddenly how scruffy they were. Dr. Shanker was hideous with his swollen face and thick, matted beard. The dinosaur-skin sack was still tightly wrapped about his rib cage. His clothes were barely recognizable as such, and would not last much longer.

Yariko was little better. Her face was dirty and streaked with sweat. Her arms were bruised, scraped, and not very clean. Her jeans were covered with dried mud and her T-shirt was no more than a bleached, gray rag. It did not conceal much, either.

"Yes?" Yariko said, smiling faintly and pretending to adjust her clothes. "You have a problem with my attire?"

"No," Julian mumbled, embarrassed. "It's just . . . you could use a bath."

Yariko smiled. "Then I'm in the right company. I've never seen a dirtier face than yours."

"What do you mean?" Julian said indignantly, putting a hand to his face. "I need a shave, but—"

"Whitney," Dr. Shanker interrupted, "find a stream and dip your head in it."

They found running water not far from the caves, behind a stand of low, twisted trees. Yariko called the trees the shower curtain, and the analogy wasn't bad. The stream tumbled over a mass of rocks just above their heads and fell in a fine spray into a rocky pool. It almost looked as though it had been arranged that way.

Julian would have loved a cold shower then and there but Dr. Shanker was hungry and insisted on dinner first. They managed to kill a dozen small lizards that were warming themselves on the rocks, trying to catch the last of the evening light. The creatures were mostly skin and brittle bones, but all together made a decent meal. Hilda contentedly crunched on the leftovers.

Dr. Shanker carefully banked the fire rather than scattering it. "Might as well keep the heat," he said, and yawned. "I can hardly keep my eyes open. That bath will have to wait until morning. You two go ahead. Watch out for predators and don't let the fire get out of hand." With that advice he staggered into the cave, Hilda right behind him.

282

Yariko and Julian sat for a while outside the cave, beside the embers of the fire. The red light of sunset faded and stars gradually appeared. After a while Yariko reached for Carl's sack. They had carried it along, not knowing what useful things might be inside. She pulled out the strange musical instrument and laid it across her lap.

"There are two ways to hold it," she murmured, gently touching the wood. "How did he play it?"

"Upright." Julian gestured to show the position.

Yariko set it upright in her lap, plucked tentatively, and smiled at the result. Then she tried an experimental chord with her left hand. "Amazing," she said. "It's tuned essentially like mine." She played a melody, a sad wandering tune, leaning forward with her ear close to the instrument, and smiling.

The last time Julian had heard music from that instrument it had brought Yariko to mind with painful longing. Now it evoked images of Carl. He listened, marveling at how they had passed down Yariko's music to their children, and they to theirs. It was the last thing he would have expected to be treasured in a world where basic survival was of such primary importance. What had it done for them, these people hunting with spears, dodging *Tyrannosaurus rex*, competing with the fierce raptors for food? But perhaps it was not surprising after all. Music was one of the most basic forms of human communication, developed long before written language existed.

After a few minutes Yariko stopped and carefully replaced the instrument in the sack.

"I have a present for you, Jules," she said. "I made it myself."

She got up and stepped carefully into the dark cave, one hand outstretched to feel along the wall. When she came out she had several of Carl's ponchos folded over her arm.

"That's not the present," she said. "Come on, I'll show you."

She took Julian's hand and led him toward the "shower," the stream that tumbled down over the cliff, where she laid the hides on the stony ground. Then she took something out of her pocket and held it out. It looked like a small stone, a smooth grayish lump. Julian prodded it suspiciously with a finger. It had a waxy feel.

"Soap," she said. "Drippings and ashes."

"That's your present?" he said, skeptically. "Homemade soap?"

"Trust me," she laughed. "You need it pretty badly." Then her voice became serious. "Jules, I missed you. All that time I was alone out there, I couldn't stop worrying. I didn't know what had happened to you. I didn't know if I'd see you again."

"I was frantic about you," Julian said, and he put his arms over her shoulders and hugged her fiercely. "Yariko. It's wonderful to have you back. And you're never going anywhere without me again."

In the near-total darkness under the bright stars they helped each other to undress. The night air was warm and still under their private cliff. They felt safe, protected, and very much alone together. Standing under the small cascade they wet each other's bodies with handfuls of water. The stream was cold and they stood close to keep warm, massaging the soap into each other's skin and hair.

Yariko gave a little cry as the soap slipped out of her hand and disappeared into the darkness. "No matter," Julian murmured, pressing his soapy body against hers. When he kissed her he tasted blood and realized that the cut on her cheek had opened. He moved his lips over her face to her mouth, and her chin, and then down her throat. The ground was cold and hard under their bare feet and Julian pushed her back until she was lying on the hides.

Sixty-five million years ago an asteroid struck the Yucatan peninsula in the southern part of the Gulf of Mexico, leaving a huge impact crater. This disaster is permanently recorded in a layer of rock that is rich in soot, dust, and debris, and also iridium, a signature element of asteroids. Heat and fiery fallout killed much of the life in the western hemisphere; dust rose to the atmosphere and produced a global blackness lasting from months to years. Ecosystems all over the earth collapsed. Half of the plant and animal genera disappeared. In the next layer of sediment, there are no dinosaur bones. The titans that had reigned for a hundred and fifty million years had vanished, and the living world was forever changed.
—*Julian Whitney,* Lectures on Cretaceous Ecology

– TWENTY-FIVE –

Julian woke at the same moment as Yariko. The cave was filled with a thin gray light. He was so used to sleeping in the open that he was disoriented at first; but then the evening came back to him. They'd taken the hides into the second cave, leaving Dr. Shanker and Hilda alone in the other.

Some sound had woken them. After a moment it came again: Hilda's barking, and then Dr. Shanker's voice coming faintly from outside, telling her to sit still and stop jumping on him. Julian smiled and propped himself up on one elbow to gaze at Yariko. Her hair was sleek and shiny after the bath, and she smelled of wood smoke. He traced her cheekbone with a finger and realized suddenly that Carl's face had seemed familiar because it had the same shape as hers.

"Yariko," he said.

"Yes?" She lifted her arms and encircled his neck, smiling. The lines and puffiness of the night before were gone and she looked absurdly young with her bare shoulders showing above the covers.

"What was this animal you said followed you?"

"The animal? The T. rex, of course."

"How long was it trailing you? What did it do?"

"It stalked us for about a week. Sometimes it would disappear for a day or two, and then it would show up again. At first it kept

285

at a distance and watched us. But then it started to come closer each time, and Hilda would go crazy. I'd have to drag her into the bushes and try to hold her muzzle to stop her from barking. I was sure we'd be eaten. It didn't attack us until that last night. I don't know what it wanted. I suppose I should be grateful, though."

"Why grateful?"

"She may have scared off something else, one night. Something that looked a little like the dromaeosaurs, but smaller. I looked up from my fire and there it was, just staring at me from about thirty feet away. I didn't even know how long it had been there. It was creepy, so silent. Then that T. rex blundered by. I only caught a glimpse of it through the trees but I felt the ground shake. It woke up Hilda and scared away the other animal. You know, the T. rex frightened me less. Isn't that strange? Maybe something so huge is a little beyond our comprehension. Carl seemed to know it, almost; as if they were friends."

Julian nodded. "He did seem to know her pretty well. They'd been together a long time. Decades, I think. She was following us, too, you know; she must have been traveling back and forth between your camp and ours. An animal like that can cover a lot of ground quickly."

"But isn't that ironic," Yariko said. "If only it could have talked. I could have sent you a message. I thought I would never see you again; and all that time, it knew perfectly well where you were. What did it want?"

"I don't know." Julian felt sad again. "She seemed drawn to people, or at least to Carl. She'd lived around his hill his whole life, you know. Maybe her whole life, too."

"You think she was domesticated?"

"No, not exactly. But she was clearly used to humans, and to finding food near them."

A voice from outside interrupted their conversation. "Anyone for breakfast?" it said, and Hilda barked.

The morning was strangely quiet, altogether different from the physical strain and danger of the previous two months. They caught small animals, mainly lizards, and gathered edible plants that grew out from the cracks in the rocks. They saw no dangerous predators in that elevated, stony terrain.

But the quiet seemed deceptive to Julian. He became nervous

as the sun climbed higher. The suspense was more difficult to take because there was no longer a physical task or goal to occupy them. He wanted to take Yariko's hand and run with her away from the caves, from any possibility of reversion, from the uncertainty and the possibility of separation or death. But instead he suppressed the panic and forced himself to do the little routine tasks that they'd worked out; cutting brush for fire, gathering water and food, carefully marking the near edge of the reversion area.

Dr. Shanker was already regaining his strength, although his limp was worse again. The bitten ankle must not have been quite healed when he strained it again carrying Carl's bier. One side of his body was badly bruised and turned a sickly brownish yellow, but his lungs seemed to be healing well and he was almost his usual active self. After breakfast he did a hundred pushups and then went off to wash his face in the stream.

He seemed a quieter person than previously, less full of himself and more considerate. Occasionally his ego would reemerge, but by and large he had improved. Julian did not think anybody could have maintained a sense of personal importance against the things they had seen and lived through.

Late in the morning they paused in the busy work to enjoy the clear sky and sun in front of the caves.

"Only one day to wait, now," Dr. Shanker commented. "Assuming of course the vault wasn't destroyed and is still set for our return. Either way, by the end of tomorrow we'll know that we're here for the rest of our lives."

"Is that what you hope?" Yariko asked.

He thought for a moment, passing his hand over his face, and then said, "No, I'd rather go back. I miss my own world. I miss my lab. Although. . . ," he paused, looking down into Hilda's panting, smiling face. "She's got ten or twelve more years in her anyway, which is about all I'd have if we stayed. . . ." Then he smiled. "Should we take a vote? Whitney?"

Julian thought about the dangers of the present world, compared with the modern world overrun by man. There were no cars in this time, no money, no overpopulation; the landscape was not burnt, clear-cut, paved, or otherwise devastated by human activity. For himself, he could almost have stayed in the Cretaceous. Part of him longed to.

But what would life be, if they stayed? Survival only, living from day to day. Their line would dwindle and die out after a few generations. In the twentieth century, he could at least make a difference in the world. He would teach, and have children. He, Julian Whitney, would make his own infinitesimal contribution to the course of human thought and evolution. Here, he and all his descendants would disappear, as if they had never been.

Julian weighed the different issues in his mind and they came to a balance.

"I'd rather stay here," he said lightly, taking hold of Yariko's hand. "I've got everything I want."

"In that case, I'll stay too," Yariko said, laughing. "You can go back, you and Hilda, and give everyone our best regards."

"I'll make sure you get half the credit for the Nobel Prize," Dr. Shanker said. Then he grew serious. "Wishes aside, we need a plan in case we don't revert. We can't spend the winter here; I know there's not much seasonality, but even now there's hardly any food and this miserable brush won't provide enough firewood. We're almost through the stack of wood inside."

"There's Carl's hill," Julian said. "It's safe enough from predators, and only an earthquake could seriously damage it. The hut's roomy enough for the four of us, and the corral could hold a small herd of animals." With his words a great longing for the hill came up in him. He wanted only to go back there, and live as Carl had lived.

"Of course," Dr. Shanker said. "After all, we built the thing, didn't we?"

"What about that T. rex?" Yariko asked. "It may have gone back there. I'd rather not face it again."

"Maybe we injured it and it won't live," Dr. Shanker said.

Julian shook his head. They had hardly scratched Corla. She might die of hunger in the desolate landscape of the hills, but Dr. Shanker certainly could not have killed her. Julian suddenly knew that he wished her well. An animal such as that should not die of anything as slow and humiliating as starvation. They owed her something: her eerie presence at the end of their journey had kept off the smaller, more dangerous predators. She had done as much as Carl had to get them here. He found himself wondering how long such creatures lived. She might be more than a century old;

she might have known the versions of them who had come here before.

"T. rex or not," Dr. Shanker said, "in two days we had better start back. Or rather, you two can. Hilda and I are planning to be out of here and back in the lab."

But stochastic processes, as Dr. Shanker would have said, or chance as most people would call it, can play tricks on everyone. Julian was cutting up brush for a fire to smoke the meat of a small dinosaur, in case they needed packable food for a journey back to Sentinel Hill, and as he worked he pondered the strange geometric diagram in the cave. He couldn't help feeling that it was important.

They already knew the day of their time window: the third day after the new moon. He and Shanker had carefully paced out the nearest regional boundary, with Yariko inside the cave shouting out the measurements from various landmarks. There was always the chance they were a little off, but as long as they were well inside the estimated boundary they should be safe. What more was there to know? Why couldn't he stop worrying?

Because, Julian's mind answered itself, you've spent two months fighting your way west against a deadline, and now you're unable to sit back and relax, knowing you're a day early. It doesn't feel right.

Yariko could be heard humming to herself as she sat near the cave skinning the animal to be smoked. Julian smiled at her domesticity: his wife, his partner, skinning dinner at the entrance to their cave, humming a little tune, a Japanese tune that Carl had played on his instrument.

He wiped his forehead with the back of his arm—he was sweating in the full sun as he tried to chop the tough, twisted brush—and looked up at the azure sky. There was no morning shade in this tree-less place. Fortunately the sun was nearly as high as it would get, and shade would soon be creeping out from the cliff face. The moon, still very thin but noticeably larger than it was yesterday, was keeping pace with the sun as it rose in the sky. Three days from the new moon . . . the moon rose later each day as it waxed and fell behind the sun, until it shone out at night when full . . . the new moon . . . the crude axe dropped from Julian's hand with a thud, scattering his careful pile of brush.

"Yariko!" he yelled, or rather shrieked, and he ran toward her, painfully wrenching his ankle in his haste.

Yariko jumped to her feet, the bloody animal dangling by one leg from her hand. "What is it? What did you see?"

"The moon!" he cried, pointing up at it. "The new moon! It *is* three days past the new moon. Today, not tomorrow. Today is the day." Julian was panting as if he'd run a mile instead of forty feet.

Yariko looked bewildered. "Are you sure? I remember seeing it two days ago, when . . . before Carl . . . I remember seeing it rise, just before the sun. It was new. In the cave it says reversion was on day three—"

"This is day three!" Julian almost shouted. "The new moon isn't the sliver we saw that morning. A 'new moon' is when none of it is visible, when the sun lights up the side away from Earth. Yesterday, when we got here, was the second day after the new moon, and today is the third."

The half-skinned animal fell to the ground. "Today . . . we don't know what time!"

"I think I can find out." Julian grabbed her arm, roughly, and pushed her away. "Go. Run. Get over there where we made the line of stones. Yell for Shanker."

"What about you?"

Julian took a deep breath to calm himself. "I'll be there soon. I think I understand that drawing. . . . SHANKER!" he yelled. "DR. SHANKER!" He ducked into the cave.

The light was poor, and he cursed himself for not remembering that. But his eyes adjusted quickly, and now that he understood it the diagram made sense, dimly seen though it was. He heard breathing behind him and turned to see Yariko's silhouette as she entered.

"I'm not going anywhere without you," she said. "Now explain this drawing."

"This is the horizon," Julian traced the wobbly curved line with his finger. "A small part of the global circumference. These lines are sun angles over the course of a day. Remember when I showed you how to stack your fists to see how high the sun was? This shows the progression of the sun in the sky, as height above the horizon. Local noon is when the sun is highest; after that, the angle decreases again. This time of year, of course, it doesn't get very high at this latitude."

"But how does it tell us the time?" Yariko squatted beside him on the dusty floor, trying to make out the vague details. One triangular wedge, bracketed by two sun angles, seemed to stand out.

"This part of the diagram is colored in," Julian went on in a rapid voice, putting his hand over the triangle Yariko had noticed. "And the little suns here, up in the sky, are much bigger. When we were here with the torch I noticed the coloring. It marks this range as our time window." He paused, frowning. "The question is, which way is east? Which is the morning sun and which the afternoon? It could make all the difference."

"East is usually on the right," Yariko said. "Isn't that the convention with maps?"

Julian put his face nearly against the rough wall as he felt around with his fingers. "There must be a note of the cardinal directions," he said. "If only I could see—"

"There! It's underneath the horizon. It's hard to see."

Julian couldn't quite make it out. "Which way is east?" he asked, almost dreading the answer.

"That way . . . right."

"Then our time window is midmorning to local noon. It's almost over."

• • • • •

2 September
5:50 AM Local Time

Earles was the only one who didn't jump when the door of the outer room slammed open, sounding like a gunshot in the predawn quiet. Two police officers rushed in.

Ridzgy's mouth fell open. "You're going to arrest those people if they reappear?" she asked, staring at Earles.

For once Bowman was quicker than his colleague. He looked from Earles to the grim-looking officers, both large men, and walked over to Ridzgy. "It's time to go, Marla," he said gently, putting a hand under her elbow.

Ridzgy looked around and saw that everyone was watching her: Earles, Mark, the two policemen; one of them was the idiot with the cigarette breath, but he didn't look like an idiot now. He looked grim, decisive. She stood. "What are you talking about?"

Earles stood also. Her casual pose was gone. "Marla Ridzgy and Claude Bowman, you are charged with conspiracy, theft, and sabotage—for a start. You have the right to remain silent. You have the right to counsel. Anything that you say. . . ."

Hann stepped forward with the handcuffs.

• • • • •

For one moment of frozen disbelief Yariko and Julian simply stared at one another. Then the adrenalin took over.

They scrambled for the cave entrance.

There was a ludicrous instant when each tried to make the other leave the cave first, but Julian won: he shoved Yariko bodily through the narrow opening. Screaming for Dr. Shanker and Hilda, they pelted over the uneven ground, past Julian's abandoned pile of brush, past the cairns, on across an open space to the line of rocks marking the boundary.

Yariko's shouts changed as she ran: her voice went up an octave and the name she yelled over and over was "Clifford!"

When he realized what it meant, Julian had to suppress a wildly irrelevant, not to mention irreverent, impulse to laugh. "Dr. Shanker!" he yelled again, forcing down the hysterical giggles. In his momentary distraction he lost his stride.

Yariko was ahead, her hair bouncing behind her; now she was over the stone line and still running, as if unable to stop.

A shout from behind made Julian turn. "Whitney! Yorko! What's going on? Have you gone crazy?"

Dr. Shanker was back near the cave, Hilda barking at his heels as she sensed the humans' excitement.

"The time window!" Julian yelled. "It's now! Run!" He started to run faster; now the line of stones was only fifty feet away. Yariko skidded to a sudden halt and ran back toward him. She paused at the stones, urging him on with her arms and eyes.

Hearing a heavy sound and curses from behind him Julian looked over his shoulder. Dr. Shanker had fallen on his knees. But even as Julian turned, Shanker was up and running again; running slowly, limpingly, on an obviously weak ankle.

"Julian!" Yariko shrieked.

Julian stood now halfway between his two companions. He took one step, one slow step, toward Dr. Shanker.

"No! Go on! Don't help me," Shanker cried fiercely. "Get in there with Yorko! Move!"

Julian sprinted the last forty feet right into Yariko's outstretched arms. She steadied him, panting.

They watched Dr. Shanker coming closer with agonizing slowness. Hilda kept pace with him; she obviously thought this was a game. "Stay there! I'm coming too," Shanker called.

"He's going to make it," Yariko said, and then the world dissolved into roaring noise and confusion.

• • • • •

2 September
6:30 AM Local Time

"I think it's ready now." The young man, eager and excited although tired, climbed out of the vault. "I've made the fine adjustments. We can power up now." He went to the computer.

Earles took a deep breath and let it out. She'd first seen this lab less than twenty-four hours ago, yet it felt like she'd been in and out of it all her life. She heartily hoped never to see it again after this morning.

Mark halted the string of commands on the IBM. "Switched on," he said. "I'll do the final settings now."

Earles followed him to the vault and watched as he crouched over the dials with his delicate tools. A few minutes went by, and she restrained the urge to ask questions; he'd warned her about vibrations, including sounds. She returned to her seat near the computer.

Then Mark emerged, triumphant, with an excited grin on his face. He left the door open behind him and went back to the computer.

"Perturbation level six," he said, finger poised over the enter key. "Here we go."

• • • • •

Julian lay as he was, eyes closed, surrounded by the harsh stench of a synthetic world. He did not dare to move. Please, he thought, please let me not be alone.

Finally he opened his eyes and looked into the dim, flickering light of a tiny room. Bits of twisted wire and shards of glass lay scattered about. There was the sharp smell of blood. He sat up with an effort, feeling the blood run down his face, and looked around, dreading what might, or rather might not, be with him. With a sigh, he reached his hand out to Yariko, and held tight.

When it was all over, the invasion by more people than he could comprehend, OSHA officials bursting into the middle of it all and sealing off the vault, packing up the computers, shouting orders in the confusion; the questions, the phone calls, the discussions, and more questions; Yariko quietly, succinctly, steadfastly repeating the partial truth of spatial translocation, without giving anything else away; the questions about Dr. Shanker and Frank, and the sad admission of loss; the ambulance, the police, the doctor, the local paper; when he could not remember his phone number, or even what that meant; when someone had covered the new bloodstain on the floor with a white cloth, and Julian with a white lab coat to hide the tattered clothes, as if his Cretaceous self were another body to be discretely hidden; when it was all over, he found himself alone, walking down the corridor, climbing the stairs and then standing on the concrete steps of the building, looking out at the world.

Maybe he expected to find the forests and intense blue sky of the Cretaceous. All he saw was the bleak gray light of a September day, the brick of the campus buildings, and the grainy black of asphalt. The noises confused him and each one seemed overly loud. Cars passed on the street, people talked, doors banged, bicycle chains clattered; all this fused together into a nightmare jumble, numbing his senses.

The vast forests that he had become accustomed to, the great inland sea, the herds of ten thousand *T. horridus* were long since gone. Dr. Shanker and Hilda had already lived out their lives and died. An entire world had passed away. Somewhere under Julian's feet, a few mineralized bones remained.

– Epilogue –

The shy-looking man stood for a moment just inside the door at the back of the lecture hall. The seats were full. It was a popular course, and seventy students had registered. They were talking in loud voices, laughing, comparing schedules. Some of them were glancing at their watches. He was only a minute late, but on the first day of class the teacher should try to make a good impression—especially if that first day had already been postponed a week because of a shotgun wedding.

A touch on his arm made Julian start. A woman stood beside him; he didn't know her. She had gray-blonde hair, muscular arms in short sleeves, and a tough look about her. Stuck in her wide belt were a VHF radio and a large bundle of keys.

"Sharon Earles, Creekbend police chief," she said, holding out a very calloused hand. "You obviously don't remember me, Mr. Whitney. We met when you reappeared in the vault."

Julian did remember her now. Of all the staring, demanding faces that surrounded him that strange morning, hers was the one he finally saw clearly. She had silenced the crowds, banished the reporters and the regulators, and let him go, alone, to recover.

"I didn't say thank you," Julian said, realizing immediately how foolish he sounded. "You . . . understood."

"Perhaps," Earles said. She was as tall as he was, he saw; he didn't have to look down into her face. "But there's a great deal I still don't understand. I want to talk to you about . . . beetles."

Julian gaped at her. "My class has already started. It's Yariko who knows the physics, anyway." His hand went involuntarily to his jeans pocket.

"Dr. Miyakara is said to be on leave of absence from the University. No one seems to know where she is." She pulled a well-handled piece of paper from her pocket and unfolded it to show a close-up photo of a brightly colored beetle. "Tell me," she said. "Where were you really, for those two months?"

Julian's fingers closed around his own precious scrap of paper, the telegram he'd received that morning from the Baja Peninsula of Mexico. He had it memorized:

"Mark arrived with copies of all files stop Lab all ours stop Vault under construction stop Shanker will be proud stop Can't wait your arrival Thanksgiving stop Love Y."

He gave Earles a sweet, boyish smile, turned, and walked down the slope of the central aisle. The noise in the room died out, and before he reached the front every eye was fixed on him.

He turned around and leaned on the podium. At the back of the room, under the recessed dimmer lights, Earles could just be seen standing where he'd left her. He gave her another smile.

Then he nodded to the class, and began to talk to them. The students were so attentive that he was able to speak in a quiet voice and still be heard.

"This course is about the ecology of the Cretaceous, the end of the Age of the Dinosaurs. The entire course will focus on one example, about which scientists know a great deal: the western part of North America, during the Maastrichtian age of the Late Cretaceous. The goal of the course is to teach you what the world might have been like at that time and place.

"What was the climate like? What were the geological features of the terrain? What kind of trees grew in the forests? What kind of plants floated in the water? If you could pick up a walking stick and hike through the jungle, what animals would you see? Opossums? Insectivores? Birds? What kind of dinosaur life? Would the jungle really be teeming with monsters ready to kill you, as you might have seen in the movies? Or would the large animals be more rare, shy of your strange scent?

"I intend to tell you a story. For anyone without a syllabus, the premise of the course is that I have just come back from a two-month safari in the Maastrichtian. I am going to describe to you, over twenty-five lectures, what I saw, what I ate, how I lived, and what the world was once like. I'd like to start, however, by asking you a question. The question is a test, in a way, of your ecological wisdom. What do you suppose was the single most common type of animal, in the North American Late Cretaceous?"

The students seemed to be paying close attention, but they were all shy of answering.

"We can take a vote," he said. "How many think it was some species of dinosaur?"

More than half the class raised a hand.

"Maybe Tyrannosaurus rex?"

Most of the hands went down, but a few stayed up.

Julian smiled. "I'm glad you'll have something to learn from this course. The most common type of animal by far, in that age and in this, was coleoptera- -the humble beetle."

– Glossary of Terms –

Alphadon: (mammal) "First tooth." A small, omnivorous marsupial, about 1 foot long. Diet: fruit, insects, and small animals. Alphadon was a tree dweller with a prehensile tail. Range: North America, Late Cretaceous.

Apatosaurus: (dinosaur; formerly brontosaurus) "Deceptive lizard." An herbivore 70 to 90 feet long, 10 to 15 feet high, weighing 30 to 35 tons. Range: North America, late Jurassic.

Ankylosaurus: (dinosaur) "Fused lizard." An herbivore 25 to 30 feet long, 4 feet tall at the hips, weighing 3 to 4 tons. Range: western United States, Late Cretaceous.

Batholith: A mass of intrusive igneous rock formed deep within the earth's crust that, due to erosion, has an exposed surface of 100 km² or greater.

Ceratopsia, ceratopsians: (dinosaur) "Horned dinosaur." Suborder of herbivorous dinosaurs from the Late Cretaceous. They had beaks and bony head frills along the back of the skull.

Champosaur: (reptile) A fish eater living from the Late Cretaceous into the Cenozoic. About 5 feet long, living in rivers and swamps. Range: North America and Europe.

Coleoptera: "Sheathed wing." Coleoptera, or beetles, make up half the known animal species on Earth. Because their outer pair of wings is a relatively hard structure, beetles fossilize better than

299

do any other insect.

Cretaceous: The last period of the Mesozoic Era, lasting from 144 to 65 million years ago.

Deinosuchus: (reptile) "Terrible crocodile." Up to 50 feet long, and the largest crocodilian known. Range: North America epicontinental sea, Late Cretaceous.

Dromaeosaur, dromaeosaurus: (dinosaur) "Fast running lizard." A carnivore about 6 feet long. Range: Alberta, Canada; Montana. Late Cretaceous. A fast-moving predator with large eyes and a sickle-like claw on each foot. Dromaeosauridae were probably the most intelligent of dinosaurs. They include the Velociraptor.

Edmontosaur: (dinosaur) "Edmonton lizard." Hadrosaur. An herbivore, 42 feet long and 10 feet tall at the hips. Range: western North America, Late Cretaceous.

Hadrosaur: (dinosaur) "Bulky lizard." Duck-billed herbivores, and the most common dinosaurs. 10 to 40 feet long. Range: North America, Europe, and Asia, Late Cretaceous. Hadrosaurs are divided into crested and noncrested types.

Hell Creek Formation: Upper (late) Cretaceous deposition in North America, to the west of the western Interior Seaway. Named for Hell Creek near Jordan, Montana; the formation occurs in regions of present-day Montana, North and South Dakota, and Wyoming.

Ichthyornis: (bird) "Fish-bird." 8-inch long, tern-like bird with toothed jaws. The first known bird with a sternum (keeled breast-bone), such as modern birds have. Range: western North America, Cretaceous.

Maastrichtian: 71 million to 65 million years ago. Named for the city Maastricht in the Netherlands, where many fossils (including the first mosasaurs) were found.

Mosasaur: (reptile) Aquatic reptiles of the Late Cretaceous, very common in the US Inland Sea. Some species reached lengths of 59 feet. The first skeleton was discovered about 1780, in the Netherlands.

Ornithischian: "Bird-hipped" dinosaurs, one of the two dinosaur orders (the other being saurischian). 2- or 4-footed herbivores, including hadrosaurs, stegasaurs, ankylosaurs, and ceratopsians.

Ornithomimus: (dinosaur) "Bird-mimic." A fast-running omnivorous dinosaur from the end of the Cretaceous; 6 to 8 feet tall.

Pachycephalosaur: (dinosaur) "Thick-headed lizard." A bipedal, herbivorous dinosaur with a rounded skull, thought to have lived in small herds.

Paleocene: The Epoch spanning from 64 to 54.8 million years ago.

Parasaurolophus: (dinosaur) A hadrosaur, bipedal herbivore, duck-billed with a hollow crest 6 feet long.

Protungulatum: (mammal) "Before Ungulate." In the order condylarths, and forerunner of the ungulates. A rat-sized placental mammal known best from the Paleocene North America, but possibly originating in the Late Cretaceous, overlapping with dinosaurs.

Pteranodon: (reptile) "Winged and Toothless." A winged reptile weighing about 30 pounds, with a 25-foot wingspan.

Purgatorius: (mammal) The earliest known primate-like animal, a rat-sized placental mammal about which little is known. Incomplete fossils are mainly from the Paleocene, although it is possible the animal evolved during the very Late Cretaceous.

Quaternary: the Period from 1.8 million years ago to the present.

Quetzalcoatlus: (reptile) The largest pterosaur, or flying reptile, with a wingspan up to 39 feet, and a neck 9 feet long. Named for the Aztec feathered god Quetzalcoatl.

Sauropods: (dinosaur) "Lizard-footed." A suborder of the saurischians, sauropods were quadruped herbivores and included the titanisaurids, the largest of all land animals.

Saurischian: (dinosaur) "Lizard-hipped." Despite the name it is the saurischians rather than the ornithischians ("bird-hipped") that evolved into birds. Saurischians were divided into the sauropods and the theropods.

Tertiary: The Period of the Cenozoic Era from 65 to 26 million years ago, between the Cretaceous and Quaternary Periods.

Theropod: (dinosaur) "Beast-footed." A suborder of the saurischians, theropods were fast, bipedal carnivores. They generally had long back legs, short forelegs for grasping prey, excellent stereoscopic vision, and long tails for balancing. It is this group that likely evolved into birds.

Titanosaur, titanosauridae: (dinosaur) The largest of the sauropods.

Triceratops horridus: (dinosaur) "Horrible three-horned face." A Late Cretaceous herbivore about 30 feet long and weighing 6 to 12 tons, probably living in large herds. A slow mover with a bony frill and large horns, Triceratops may have been similar to the modern rhinoceros, charging at its enemies rather than fleeing. Triceratops is noted for having one of the largest skulls of any animal; at 10 feet long, the skull was almost a third as long as the body.

Troodon: (dinosaur) "Wounding tooth." A Late Cretaceous, human-sized, bipedal carnivore (theropod), weighing about 100 pounds. Troodon was probably a fast runner, with excellent vision and hearing, and may have been among the most intelligent of dinosaurs. It was named for its distinctly serrated teeth.

Tyrannosaurus rex: (dinosaur) "Tyrant lizard king." One of the largest of the theropods, T. rex lived between 85 million and 65 million years ago. Fossils have been found in western North America and Mongolia.

Velociraptor: (dinosaur) "Speedy-thief." Theropod (carnivorous); one of the dromaeosaurids, found in Mongolia. Had a sickle-shaped, retractable claw on each foot.

About the Authors

Lisa M. Graziano, Ph.D., is a freelance editor and writer. A former professor of Oceanography at Sea Education Association in Woods Hole, Mass., she divides her time between editing, research at sea, and writing. Lisa lives on Cape Cod with her family.

Michael S. A. Graziano, Ph.D., is a professor of Neuroscience at Princeton University. When not doing research Michael spends his time with music and writing. He lives in Princeton, N.J., with his family.

About the Type

This book was set in ITC Galiard, an adaptation of Matthew Carter's 1978 phototype design for Mergenthaler. Galliard was modeled on the work of Robert Granjon, a sixteenth-century letter cutter, whose typefaces are renowned for their beauty and legibility. ITC Galliard is a notable typeface for text.

Designed by John Taylor-Convery
Composed at JTC Imagineering, Santa Maria, CA